CAREERS USING
GEOGRAPHY

Patrick Talbot

**KOGAN
PAGE**

First published in 2000

Kogan Page Limited
120 Pentonville Road
London N1 9JN

© Kogan Page, 2000

British Library Cataloguing in Publication Data

A CIP record for this book is available from the British Library

ISBN 0 7494 3069 9

Typeset by JS Typesetting, Wellingborough, Northants
Printed and bound in Great Britain by Clays Ltd, St Ives plc

Contents

Contents

Acknowledgements

The author wishes to thank the following:

Sue, Sarah and Charlotte for their patience; contributors of case studies for taking the time to provide information and advice; and all at the Royal Geographical Society, and Hilary Arnold, for a lot of useful help over the last few years.

Patrick Talbot is Head of Careers and teaches geography at Hampton School, Middlesex.

Introduction: Geography and the world of work

Is a career using geography for you?

Are you interested in a subject which:

☐ allows you to remain flexible about your choice of career, for geographers are employed in a very wide variety of jobs;
☐ at the same time, can lead to employment in specifically related areas like teaching, planning, mapping and the environment;
☐ helps you to develop a wide range of skills widely recognized by employers as appropriate for working in a business environment, like problem solving, team working and analytical powers;
☐ enables you to use a wide range of computer skills, including state-of-the-art technology like digital mapping;
☐ is recognized as a traditional academic subject of study;
☐ allows you to develop your studies from a broad base to specialize increasingly in topics that appeal to you, like the environment or development;
☐ gives you a greater understanding of the world around you;
☐ is the only major academic discipline to combine fostering a sense of place, spatial awareness and study over a variety of scales, from local to global;
☐ allows you to combine the practical approach of fieldwork with rigorous scientific methodology and traditional research techniques; and above all
☐ is a most enjoyable subject?

If you answered 'yes' to most of these questions, then read on...

The purpose of this book is to suggest that there are many jobs that are directly linked to the study of geography. This may be through the body of knowledge itself, or through the skills the subject should develop in its students. Wherever the term 'geographer' is used, it refers to anyone who is studying or has studied the subject of geography. A wide variety of readership may browse through this book. You may just be starting your GCSE or your Scottish Standard Grade. You may be doing it for A-level or Scottish Higher. You may be taking another route but are still interested in finding out what geographical careers there are. You could be doing a degree in the subject. The book is a provider of career ideas, and that can be appropriate to all age groups.

Geography has been a school subject for over a hundred years. In that period it has evolved from what began as an essentially place-based gazetteer of facts and figures into a mature subject offering insights into an understanding of the world around us. Indeed, a measure of its successful evolution is its popularity. In 1998, it was the most popular GCSE option and the fourth most popular A-level by recorded entry. It is also studied by some 16,000 undergraduate students every year in higher education in 98 university departments.

The subject provides unique elements for the school curriculum – like a study of place, and of patterns and processes on the earth's surface, and looking at how people interact with their environment. Geography also gives students a fieldwork experience; and where would our map reading be without the subject? In addition, geography contributes to a common battery of skills and competencies in the classroom, like literacy and numeracy.

The subject has an undoubted breadth of study and it can be pursued in increasing depth through following any syllabus. It is certainly an everyday subject, and it is unlikely that you can read a newspaper or watch a news summary without making an obvious connection to the study of geography in the classroom. No surprise then to find that the most instinctive response to why students study the subject is its interest value.

Nevertheless, it is fairly rare for job advertisements for anything other than geography teachers to ask for a qualification in the subject. Keeping a watchful eye on newspaper adverts over a six-month period did reveal a requirement of a geography degree for a Geographical

Information Systems assistant ('expertise ideally gained through a Geography degree with a socio-economic basis'), an information officer in a local government planning, transportation and building surveying service ('have a degree or equivalent in Geography, Planning or a related discipline') and a senior researcher on employment and the local economy in a county council ('an appropriate degree – Economics, Geography, Town Planning'). More generally, 'an interest in geography' was useful for an applicant for a tour operator information co-ordinator, 'a good geographical knowledge' was 'desirable' for a travel guide researcher/compiler and 'necessary' for a business travel consultant, while travel consultants for an independent travel company needed 'excellent geographical awareness'. To be fair, however, many job adverts do not actually stipulate the subject of an applicant's qualification.

The width and variation of the geography curriculum are in themselves a major source of job ideas. Consider some of the elements that may make up a geography course, and it is relatively simple to suggest career equivalents, as shown in the following box.

Geography course content in relation to 50 specialist careers

Tectonic processes   Geologist, geophysicist, seismologist, vulcanologist.

Geomorphological processes   Hydrologist, hydrographic surveyor, glaciologist, geomorphologist, geologist.

Weather and climate   Meteorologist, television weather presenter, climatologist.

Ecosystems and environment Botanist, pedologist, estate manager, landscape architect, conservationist, forestry officer, countryside officer, National Park officer, worker for: English Nature, Scottish Natural Heritage, Countryside Commission for Wales, Department of the Environment or Countryside Agency, field studies officer, organizer for local schemes:

Groundwork Foundation or British Trust for Conservation Volunteers.

Population   Demographer, census officer, researcher for a population agency, housing officer, social worker, immigration officer.

Settlement   Town planner, civil engineer, land surveyor, town centre manager, site analyst for a major retailer.

Economic activities   Primary – farmer, forestry officer, mining surveyor; Secondary – manager in a manufacturing industry: cars, breweries; Tertiary – recreation and leisure centre manager, tourism officer, City worker, civil servant, traffic engineer, air traffic controller, transport planner.

Development   Worker for: the Diplomatic Service, British Council or Directorate of Overseas Survey, charity officer.

Source: Talbot, P (1999) A world of opportunities, in Promoting Geography in Schools, *ed A Kent, p 47, Geographical Association*

Details of these suggested career areas, together with many others, will be discussed in the following chapters. Although it is a perfectly logical list of job areas, this can be seen as a rather theoretical approach to career planning. Not all geographers go on to follow a geographical career, and the subject should not be seen as an entry to an exclusive range of jobs.

So what jobs do geographers actually do? The only really clear-cut evidence exists for graduates. Even then, the statistics should be interpreted with caution. They tend to be based on the first job that the graduate does. However, research suggests that graduates may not embark on a meaningful career path until three or four years after graduation. That said though, the statistics are revealing, as indicated in Table 1.1.

The table shows that a quarter of 1996 geography graduates stayed on in higher education for further study and training, which includes

Table 1.1   *What geography graduates did in 1996*

First destinations	
Employment in the UK and abroad	60%
Further study and training	25%
Not available for employment, unemployed or seeking employment	15%
Kinds of work	
Financial, clerical and secretarial	32%
Managerial and administrative	22%
Professional and technical	18%
Other	28%

Source: UCAS/CSU (1998) What Do Graduates Do?
Sources used in the original compilation of the data:
   HESA Individualised Student Record 1995/6, reference July 1996
   HESA First Destinations Supplement 1995/6, reference December 1996
Copyright Higher Education Statistics Agency Limited, 1996
Reproduced by permission of the Higher Education Statistics Agency Limited
HESA cannot accept responsibility for any inferences or conclusions derived from the data by third parties.

postgraduate teacher training courses and research for Masters degrees and PhDs. Teacher training provides the obvious pathway for graduates to teach geography, and at present no tuition fees are payable to enter a geography PGCE course. Further degrees (mainly Masters courses or PhDs) rely upon funding being available from specific university departments and government organizations. Some graduates see an MA or an MSc as a more exclusive qualification in an era of mass higher education or as a pathway to a more specialized career. Over the following chapters, we will see that this is a common way for generalist geographers to begin to specialize.

Most of the remaining three-quarters of geography graduates went into employment. Over a half of that category went into financial, clerical and secretarial jobs, and into management and administration. These are broad statistics but they do illustrate that most geography graduates do not enter careers that seem directly relevant to their degree. However, it is difficult to be precise about what proportion

Table 1.2 *Career categories for a sample of geography graduates*

BAL	Royal Air Force	helicopter pilot
CAB	Civil service	administration trainee, DFID
CAG	Local government	principal trading standards officer
CAS	Personnel	human resources director
CAV	Computer work	IT project manager
		IT consultant and associate partner
		marketing systems manager
COD	Management consultancy	external business consultant
		management consultant
FAB	Education	geography teacher, secondary school
		geography teacher, secondary school
		deputy headteacher, primary school
FAC	Journalism	foreign correspondent, daily newspaper
FAM	Religion and church work	curate
GAJ	Sport and recreation management	sport development manager
GAL	Radio, TV, films and video work	independent TV producer
		independent TV researcher
		BBC radio and TV producer
KEM	Charities and voluntary work	charity programme funding officer
NAB	Accountancy	trainee accountant
NAG	Insurance	insurance business manager
NAK	Financial services	finance manager
		finance and administration employee
NAM	Commodity markets	equity sales employee
NAN	Pensions management, etc	property investment manager
OG	Public relations/ selling	company representative
OM	Selling/buying	director of trading company

QOL	Earth and environmental sciences	hydrocarbon logging engineer field operations manager, oil company
UL	Landscape architecture	graduate landscape architect
UM	Surveying	estates surveyor
US	Town and country planning	principal planning officer development manager, overseas private practice
UT	Mapping and charting/cartography	GIS manager GIS catastrophe management consultant
YAB	Air transport	commercial pilot trainee air traffic controller
YAD	Road transport	haulage operations manager

Source: Survey of Hampton School geography graduates undertaken in June 1998 by Patrick Talbot

are involved in doing a 'geographical' job, for such statistics focus upon job titles rather than the career sectors that are easier to match to degree discipline. In this respect, smaller qualitative surveys provide more meaningful results in terms of how geographical the graduate jobs are. One such example was provided by tracking former pupils from the author's school who went on to complete degrees in geography, as demonstrated in Table 1.2.

What is particularly impressive about the results shown in the table is the wide range of careers that these geographers have entered. The inclusion of the CLCI abbreviations reinforces that width. The Careers Library Classification Index is used in all careers rooms, and information about various jobs may be found under the system. The careers in the sample cover a number of the abbreviations used. It seems that geographers do anything and everything. In fact, many of them went into areas that are traditionally popular with geographers – planning, education and surveying. Many did not! In this book, Chapters 3–11 concentrate upon the more obvious geography-related careers, while Chapters 12 and 13 look at media and commercial jobs respectively, and why geographers have the skills to do them. In

Chapter 2, we turn to a consideration of the skills that geographers acquire.

Useful addresses

The key organizations that represent geography are:

The Royal Geographical Society
(with the Institute of British Geographers)
1 Kensington Gore
London SW7 2AR
Tel: (020) 7591 3000
Fax: (020) 7591 3001

The Royal Scottish Geographical Society
40 George Street
Glasgow G1 1QE
Tel: (0141) 552 3330
Fax: (0141) 552 3331

The Geographical Association
160 Solly Street
Sheffield S1 4BF
Tel: (0114) 296 0088
Fax: (0114) 296 7176

2 The skills you develop through geography

The unique and shared contributions of geography to the curriculum have already been outlined in Chapter 1. Now we shall consider what skills employers generally look for, and how studying geography should help you to develop them. Employers say they want to recruit people who have a variety of skills:

◆ **communications skills:** vital in the world of work, where few people operate alone;
◆ **teamwork skills:** most companies and organizations structure their tasks in small group teams;
◆ **flexibility:** a skill necessitated by the complexities of working with a variety of individuals and groups, and at differing scales of activity;
◆ **self-management:** a necessary personal quality to progress in business;
◆ **powers of analysis:** important when considering day-to-day issues;
◆ **problem solving:** necessary from time to time in the workplace;
◆ **decision-making:** once you have considered the merits of various courses of action, you need to decide what should be done.

Geography helps develop these qualities in various ways:

◆ **Communications skills:** Literacy and numeracy lie behind written and oral skills, of course. Note taking, paragraph answers, essays and project work illustrate the variety of written demands in geography courses. Oral skills are developed in classroom discussions and presentations, and later on in tutorial groups. Technological skills are honed by information searches, analysis and perhaps the use of the word processor in producing an individual study.

◆ **Teamwork skills:** Fieldwork is an essential component of courses and an ideal setting in which to develop teamwork and indeed leadership skills. Decision-making exercises in the classroom may involve small group work.

◆ **Flexibility:** Geographical studies inevitably involve coping with a wide range of subjects, materials and scales. It is particularly the overview or synoptic perspective possessed by good geographers that is appropriate to working in a business environment.

◆ **Self-management:** An individual study involves a sequence of stages from planning, through collecting data, to interpretation and writing up. Deadlines have to be met!

◆ **Powers of analysis:** Questions in geography inevitably ask you to 'describe' and to 'explain' phenomena. You are applying analysis in answering the latter.

◆ **Problem solving:** This is invariably the end result of using your powers of analysis in the subject. It may be done formally, through setting up and testing a hypothesis, or informally, in explaining spatial patterns.

◆ **Decision-making:** Relevant exercises frequently appear in geography syllabuses. Decisions are made through analysing the nature of a problem and deciding which course of action is most appropriate.

In my detailed study of geography graduates, individuals were asked what skills they reckoned to have developed through studies in the subject. The most common responses referred to **analysis** and to aspects of study skills (**organization, enquiry, research** and **investigation**). Further common answers of a general nature involved **working in a group** through course elements like fieldwork, **giving presentations** and **report writing**. Rather more individual suggestions were **prioritization** and **time management**, as well as the more obvious **numeracy** and **literacy**. Not surprisingly, there were also more specifically geographical suggestions: **cartographic skills, working at a variety of scales** and **spatial awareness**. Much of what they said bears out the skill set previously outlined.

The sample of nearly 40 graduates were also asked if they felt that geography was still relevant to their working life, and if so, how? Here is a range of answers:

◆ *The use of computers is very useful in my present job; presentation skills are also used at work; the skill to write reports clearly and concisely was developed at university* (an accountant).

◆ *As I have moved into documentary film making, my degree has helped enormously. My recent documentary looked at housing needs in the city superimposed on to themes of racial tension, poverty, drug abuse, exclusion from school and demographic divisions* (a television producer).

◆ *An awareness of spatial patterns in the built environment, computer skills and environmental assessment* (a landscape architect).

◆ *Courses on regional development and urban regeneration greatly influenced my ministry in inner-city Newcastle. Courses on humanistic and perceptual geography have informed my later thinking and research on the spirituality of place* (a curate).

◆ *The use of diagrams like flowcharts to present complex scenarios for pensions* (a human resources director).

◆ *Of extreme importance in my job is the forecasting of meteorological phenomena* (an airline pilot).

◆ *Map interpretation to appreciate what landscape actually looks like, which is helpful for low-level navigation* (helicopter pilot).

◆ *Proposal writing every day and taking an analytical approach to problems. Geography gave me the ability to understand people and 'read between the lines', important when you are managing a workforce* (IT consultant).

The responses are many and varied. Individuals have taken different elements of geography with them, in many diverging directions.

The world of work no longer guarantees people a 'job for life' and many have found their career security unpredictable in recent years. However, predictable change and career flexibility are said to be increasingly important elements in the job market. What is clear about the responses from graduates is that, through geography, they see themselves as having an enviably broad combination of transferable skills. These can be taken from one job to another as people move along their career paths, with the formal geography they did becoming a distant memory! Geographers are clearly equipped for the 21st century.

> **Top Tips for developing your employability as a geographer**
>
> ◆ If you enjoy geography and you are an able student, study the subject! You are most likely to obtain your best results in a subject that you enjoy and are good at.
> ◆ Make yourself aware of the work skills that geography helps to develop.
> ◆ Be prepared to illustrate how you have developed these work skills in writing an application or CV, and when attending an interview. For instance, completion of a fieldwork project will probably have developed your teamwork skills, problem setting and solving, numeracy and literacy, and you may well have word processed the text and used computer programs to draw the graphs.
> ◆ Focus in on areas of the subject that you particularly enjoy. Which are the relevant employment areas? If some of these jobs sound appealing, find out more about them from sources of careers material.
> ◆ Write to likely companies and organizations for work experience. Take the initiative yourself. If you can obtain experience in the world of work, it will provide a valuable insight into an employment sector. Talk to people and ask questions while you are there.

Further reading and information

The Royal Geographical Society has produced a pamphlet, *Exploring the World of Work: Geography and Careers*. This can be viewed on their Web site www.rgs.org. In addition, the Society holds an annual symposium on 'Careers for Geographers' every summer term. It is designed for sixth-formers in particular, and provides a range of ideas on choosing a career. A number of geography graduates talk about how they have used the subject in their jobs. Contact the society for details.

Most issues of the *Geographical Magazine* since April 1995 have included case studies of people in careers with a geographical content in the series 'Geojobs', which has now evolved into 'Geopeople'.

The Geographical Association has produced *Going Places: A geography careers resource pack*, which includes a students' booklet with ideas for possible careers. School and college geography departments should have this pack.

If your careers library carries either the KeyCLIPS or the Independent Schools Careers Organisation information leaflets, look out for KeyCLIPS leaflet QOL5, *Careers using Geography* or ISCO information sheet 57, *Careers with Geography*. Also, see if your careers library has *Working in Geography*, Careers and Occupational Information Centre (COIC). It contains a number of case studies of people in geographical jobs.

3 Education: Teaching geography

'Everyone remembers a good teacher' runs the headline of a national media campaign. It is a reminder to us all of the importance of education, and the key role that educators play in that learning process for individuals. All but a few have been through the education system, and the working environment of a school is a familiar one. The perspective of the teacher is a very different one to that of the pupil, however! Few pupils will appreciate the sheer volume of work that goes into teaching a class well.

The subject of geography in schools makes the major contribution to developing children's understanding of place and of the world around them through the teaching of unique skills like mapwork. Together with other subjects, it uses language, numeracy and ICT skills, whilst overlapping in content in the coverage of physical themes like weather (with physics) and human themes like development (with religious education). Geography is truly cross-curricular in skills and knowledge. Exciting approaches like fieldwork and enquiry are central to teaching the subject, and the case studies used to illustrate key elements are continually changing. No wonder the subject is popular in school option numbers!

Nevertheless, teaching as a profession has suffered recruitment problems in recent years. Applications to secondary ITT (that is, Initial Teacher Training courses to train secondary school teachers) fell by 6 per cent between 1997 and 1998. Geography itself is certainly caught in this decline. Applications to postgraduate secondary teaching courses (the Post Graduate Certificate of Education) in geography fell by 10 per cent in the same period. Geography is now officially a shortage subject in teacher training, although this will take a little more time to filter into secondary school recruitment.

The structure of teaching

There are 400,000 teaching jobs in the maintained sector, which caters for 90 per cent of children and covers the age range 3–19. The other 10 per cent of children are privately educated, largely in independent schools. The education system takes an individual through a series of establishments, from nursery into primary school and then secondary school. Education is compulsory to the age of 16. In England and Wales, the system is separate from that in Scotland.

In England and Wales, education is more prescribed. A child's progress is divided into Key Stages and regularly assessed, and teaching is focused upon the National Curriculum. Primary schooling covers Key Stages 1 and 2, with assessments measured at the end of each stage. Secondary education is similarly subdivided into two: Key Stage 3, after which is another set of assessments at age 14, and Key Stage 4, culminating in GCSEs or NVQs. Meanwhile in Scotland, pupils take Scottish Certificate of Education (SCE), Standard Grades or Scottish Vocational Education Council (Scotvec) modules. In the National Curriculum in England and Wales, core subjects have to be taught – maths, English and science, and Welsh in Welsh-speaking schools in Wales. Art, design and technology, geography, history, information technology, modern foreign languages, music, physical education and Welsh in non-Welsh speaking schools in Wales are foundation subjects. All pupils also study religious education. In Scotland, primary education must cover five main areas: environmental studies, expressive arts, language, maths and religious, social and moral education. During secondary education a wider range of subjects is studied.

In all parts of the United Kingdom, students who stay on in the system will continue into the sixth form in some schools, or move on to sixth form or tertiary college. In England and Wales, students study for A-levels, AS-levels, NVQs or a mix of them. In Scotland, students achieve SCE Higher Grades or the Certificate of Sixth Year Studies.

Those who wish to continue beyond the age of 18 stay on in tertiary college or go on to university or college, ie into higher education. There are 60,000 lecturing jobs in over 100 establishments, and intense competition for vacancies.

The jobs

Essentially, teachers choose to teach in the primary or secondary sector, or to lecture in the tertiary sector. Primary school teachers will tend to cover the whole range of required subjects with their class, including geography. It is in secondary schools and sixth form colleges that specialist geography teaching will be focused. Classroom contact time forms the pivot of the job, and establishing a good working relationship with pupils is fundamental to creating a suitable learning environment.

The various public examination syllabuses dictate much of the content. It is an advantage to offer a second, subsidiary subject, and an interest in some aspect of extra-curricular activity can only help you to secure your first teaching post. Indeed, it would be a mistake to think that you only need to teach your subject. The ultimate goal for a geography teacher may be to become a head of department. Beyond that, promotion into senior management in a school or college will inevitably lessen the geography teaching load.

Lecturing in higher education involves very different responsibilities to that of school teaching. Students are prepared for a wide range of first and higher level degrees. Lectures given to large groups of students, rather than lessons, are the basic units of teaching. In geography there may be more practical sessions, especially small group tutorials. There are opportunities to lead fieldwork, in the locality or in more exotic destinations. As well as designing, preparing and teaching courses, lecturers are required to carry on research. In older universities, the emphasis is still on research while in new universities the balance tilts towards teaching. However, it is in research that you will make your name, and get on in the academic world. Your expertise may be called upon on a consultancy basis by companies and organizations, or in the media.

There are a number of specialist teaching posts. The armed forces employ some people in a teaching capacity. Special needs teaching and teaching English as a foreign language (TEFL) are very different in context and demands from most teaching. Some teachers participate in various Voluntary Service Overseas schemes, most probably at the start or at the end of their teaching careers. Teacher or lecturer swaps can also offer the experience of teaching abroad. However, the most obvious specialism for a geographer is to work at a field centre, where the subject can be taught to visiting school groups to A-level and, as you gain experience, you may be expected to supplement the tutoring of university groups.

Allied professions

Nursery nurse, social worker, educational welfare officer, youth and community worker, educational psychologist, outdoor education leader, careers officer.

Personal qualities

A good working knowledge of your subject area is vital, especially for your confidence. It is certainly a sound maxim that you don't know your subject until you've taught it! Beyond that, establishing rapport with a class emerges through that imprecise mix of patience, tolerance, understanding, firmness, good organizational skill and sheer hard work. Importantly, that relationship stems from the tremendous amount of responsibility a teacher is given from day one. It is no 9 till 4 job, as it requires large additional amounts of time for preparation and marking. Long holidays are some compensation for the sheer demands of a normal school week. Teaching can be the ultimately satisfying job, as you appreciate the development of an individual or teaching group under your guidance.

Getting started

To become a primary or secondary teacher, you will usually need a degree. Most directly, this will be a Bachelor of Education (BEd) degree or a Bachelor of Arts or Science with Qualified Teacher Status. These are four-year full-time courses. Many people who are considering teaching choose to remain more flexible by studying their subject of interest first, after which they complete a one-year Post Graduate Certificate of Education (PGCE). To enter a PGCE course in geography in England and Wales, you need a degree in the subject, plus GCSEs in English language and maths. In Scotland, a degree plus H Grade English is necessary, and the degree should contain passes in at least two Teacher Subject Quality Courses in geography. These are the part of a degree course relating to the subject as taught in Scottish secondary schools. From 2000/01 candidates will require at least three TSQCs.

It is also possible for graduates to combine employment in a school with training that leads to Qualified Teacher Status in England and Wales. This is the School Centred Initial Teacher Training (SCITT) scheme. The Registered Teacher Programme is organized along similar lines for those with an HND qualification. In the state system, a satisfactory first year in teaching as an NQT (Newly Qualified Teacher) will complete your qualification. In Scotland, the first two years are probationary.

In the higher education sector, no formal teaching qualification is required. Full-time university posts will usually require a proven academic ability, usually a PhD. This also is desirable but not essential for work in a college of education. There will be in-house induction and training schemes to help develop communication skills.

Top Tips for getting into teaching

- ◆ Appreciate that teaching is about dealing with large numbers of pupils at the same time. Don't be fooled by the ease with which you have coped with baby sitting or with brothers and sisters, nephews and nieces.
- ◆ If you wish to consider teaching, see how you cope with dealing with children en masse. Take part in play schemes, youth organizations and summer activity holidays.
- ◆ If you are an undergraduate considering teaching, look out for 'taster' courses offered by teacher training providers and local education authorities. They usually last about three days.
- ◆ More generally, go back to your old school or college. Chat to appropriate subject teachers about the profession, and ask if you can observe some lessons.

Case Study

A classroom teacher

Sam has taught geography at a state secondary school in Surrey for four years. When she was at school herself, she talked to a relative who was a teacher and first began to show interest in the career. She completed two weeks of work experience in a secondary school, and this convinced her that she wanted to go into teaching. Sam preferred to keep her options open when she went to university locally, however, and she chose a BSc course including geography. She maintained an interest in wishing to teach so, when she completed her degree, she went on to do a one-year PGCE course.

A year later, she still didn't feel that she wanted to settle into a permanent teaching job straight away and planned to go to Hong Kong to do some supply teaching. However, she saw an advertisement for a geography teacher close to home, and applied.

'The minute that I set foot in the school I knew that I wanted the job,' Sam remembers of the day of her interview. She was offered the job, and has been happy there ever since. She loves the interaction with the children, and had the responsibility of being a tutor from day one.

'As we have a vertical tutor group system, I have responsibility for about 30 pupils from Year 7 to Year 11 (aged 11 to 16). It's very rewarding work with most of them, hard work with a few!' she thinks. As far as teaching geography is concerned, Sam loves communicating the subject matter. 'I get a real kick when I hear them discussing what we have just covered as they leave the classroom.'

Sam would like to extend the range of her teaching to include A-levels or GNVQs. She would like to be a head of geography eventually but she realizes that she may have to move on to another school for the opportunity.

Case Study

A teacher in outdoor education

Rachel found it difficult to decide what to study at A-level since she had enjoyed and achieved highly in most subjects. Whilst walking through the exhilarating glaciated scenery of the Alps just after her GCSE exams, she was inspired to make geography her main subject, and she went on to study it at A-level and university. As she had hoped, the broad range of skills required for her degree course made it both challenging and *rewarding. She chose to study geography still further by doing research for a PhD, which was based at another university. The research concerned changes in water quality during storm hydrographs. The field centre at which she now works is an important centre for such hydrological research and she was based there for her field sampling, which was by necessity almost entirely during rain events!*

As the PhD neared completion, Rachel decided to look for a job in education, as she had particularly enjoyed teaching whilst doing her doctorate. She wrote an open letter of application to the head office of the Field Studies Council, and was invited to visit their centre adjacent to the head office. This visit confirmed her convictions to get a job in teaching in a field centre and she was delighted to be invited to a central interview when posts arose at three FSC centres. She was offered the job at the centre where she had conducted her research and started work there in October 1997.

Rachel's job is to plan and deliver field courses for geography students. A typical course is between four and six days for A-level students, although the centre also delivers courses for other school groups and universities. She believes that fieldwork is a vital part of any geographical course and that it is an honour to lead students through that experience. She also feels that the broader benefits of fieldwork for personal development and environmental awareness are important. The centre is situated in beautiful surroundings, which include spectacular coasts and Dartmoor National Park.

Useful addresses

England and Wales
The Graduate Teacher Training Registry
Rose Hill
New Barn Lane
Cheltenham
Glos GL52 3LZ
Tel: (01242) 223 707

The Teacher Training Agency publishes a range of free publications obtainable from:

Teacher Training Information Line
PO Box 3210
Chelmsford CM1 3WA
Tel: (01245) 454454
Fax: (01245) 261668
E-mail: teaching@ttainfo.demon.co.uk
Web site: www.teach.org.uk

The Teaching as a Career Unit (TASC) Publicity Unit
6th Floor, Sanctuary Buildings
Great Smith Street
London SW1P 3BT
Tel: (020) 7925 5880/5882

The Welsh Office
Education Department
FHE1, 3rd Floor, Companies House
Crown Way
Cardiff CF4 3UT
Tel: (01222) 388588

Northern Ireland
Department for Education in Northern Ireland
Balloo Road
Bangor
County Down BT19 7PR
Tel: (01247) 279537

Scotland
The General Teaching Council for Scotland
Clerwood House
96 Clermiston Road
Edinburgh EH12 6UT
Tel: (0131) 314 6000
Web site: www.gtcs.org.uk

Independent schools
The Independent Schools Information Service (ISIS)
56 Buckingham Gate
London SW1E 6AG
Tel: (020) 7630 8793

Higher Education
The Association of University Teachers (AUT)
Egmont House
25 Tavistock Place
London WC1H 9UT
Tel: (020) 7670 9700
Web site: www.aut.org.uk

Further reading and information

Taylor, F (1997) *Careers in Teaching*, Kogan Page, London
Riley, J (1997) *A Guide to Getting into Teaching*, Trotman, Richmond
Just the Job! Teaching (1997) Hodder & Stoughton, London
Working in Teaching (1997) COIC

The Times Educational Supplement is published every Friday, and is available in most libraries. It is the prime source of advertised teaching jobs in the primary and secondary sectors.

4 Planning

What impacts might green belts have on the future use of land in the local area? What are the relative advantages of greenfield and brownfield sites? Should a proposed superstore be given planning permission?

These are common themes discussed in geography lessons in the classroom. They also happen to be the sort of issues that confront planners every day. Consequently, elements of human geography, and indeed some specific aspects of physical geography too, provide an excellent background to planning as a career. Planners make a central contribution to debates and consequent decisions about housing, conservation, economic development, environmental education, recreation, sport, tourism, transportation and community development. This catalogue could easily be mistaken for the themes that make up a geography syllabus!

Moreover, the role of the planner in forecasting and managing change within a complex modern society includes an appreciation of a broader and long-term perspective. Here, the synoptic view of the geographer is a particularly appropriate skill. No wonder that geography is recognized as a relevant degree by the Royal Town Planning Institute, and that many geographers go on to enter planning via a postgraduate course.

The structure of planning

Town and country planning, to give it its full title, is mainly concerned with finding the best use for land and controlling its development to meet the needs of all groups in our society. The natural scales at which we can best plan land use – local, regional and national – do of course overlap. Local issues are rarely merely local in their implications, and one area of the country cannot be allowed to carry out change in isolation. Hence came the necessity for a national planning system. It essentially grew out of the 1947 Town and Country Planning Act, which created development plans and planning applications.

About two-thirds of planners work within this system, which is pyramidal in structure. The national level offers fewest jobs, helping to produce broad policies and monitor those policies and proposals when enacted at local level. Meanwhile, the majority of employees work at the local level, preparing plans or responding to guidelines set nationally.

The remaining third of planners are either employed by large firms, transport agencies, tourist authorities and environmental organizations, or they are in private consultancy, hiring out their expertise on short-term contracts to other companies and organizations. Although the job market for planners is quite static, the private sector is expanding.

The majority of planners will seek the professional status of membership of the Royal Town Planning Institute, which necessitates completing one of the academic routes referred to later in this chapter in 'Getting started', plus two years professional work directly related to town planning.

The jobs

Central government planners
Planners at the national level work for the government, and are civil servants. Most work for the Department of the Environment, Transport and the Regions (DETR), or for the Scottish, Welsh or Northern Ireland offices. Here, they help determine national needs and hence policies, they issue circulars and planning policy guidance notes on specific topics such as transport and shopping centre location, and they adjudicate in disputes and appeals.

Local government planners
Central government determines policies, which reach fruition at county level, where councils prepare a structure plan, and at district level, where analysing local needs within wider-set constraints leads to determination of a local plan. Unitary authorities in the conurbations combine the two scales in unitary development plans. Hence, planners working for local authorities are employed by county, district or unitary authorities.

The other local responsibility is development control, generally examining planning applications to consider whether planning permission should be granted. This can be for anything from a small proposed extension to a private house to a major retail or industrial complex.

Many local authorities have larger, more specialist departments dealing with elements like regeneration and development, or more specialist expertise like historic buildings or tree protection.

Private sector planners
Such are the complexities of planning that private companies and organizations increasingly seek the services of professional advisers. Planners may be employed by such companies but more commonly they work in consultancy companies who undertake contracts for a wide range of organizations. The RTPI claim that up to 20 per cent of its corporate membership work in consultancy for over 1,250 firms. Most sensibly, planners working in private practice have probably gained early experience in local government. The fact that consultants may be completing short contracts for local authorities indicates the opportunities to move between the public and the private sector. Working abroad is also more likely working in a consultancy.

Planning technicians
Planners are supported by technically competent staff who are responsible for efficient storage of records. They help research, collate and analyse data for planners, and play a very significant practical role in producing a wide range of reports and other publications.

Allied professions

Architect, civil engineer, surveyor.

Personal qualities

Planning is a controversial sector: interested companies or organizations usually have a definite perspective and members of the public invariably have a definite opinion – remember the NIMBY (not in my backyard) from geography lessons? As a planner you will spend much of your time liaising with interested groups. Your analysis of circumstances has to reflect a variety of opinion but also must contain good objective statistical evidence. Planning work is likely to involve balancing on-site visits, office-based work and a wide range of business meetings – in office hours and outside! Basic survey work, research and analysis are necessary to provide evidence for the presentation of proposals and of reactions, both in written reports and at public meetings. You need to communicate well, to be a good listener but to be prepared to make decisions based upon the body of evidence, and to work well in a team – work is usually organized in project teams. The leadership of such teams is taken by a senior planner but few such positions are available.

Getting started

The easiest route into planning is via a degree in town planning, a four-year course. If you choose to study geography instead, you will be well advised to 'convert' to planning via a postgraduate diploma. This you can do because geography is recognized as a relevant degree, although the diploma course will take two years full time, and consequently it will take a year longer in total to achieve equivalent academic status to that of a planning graduate. Studying geography and then 'converting' does have the advantage of keeping you more flexible for longer. Or, to put it another way, you can put off that final commitment to career direction for longer! The planning diploma can also be studied on a three-year part-time basis through day release at the discretion of your employer. This entails getting a planning job on the strength of doing a geography degree first, of course.

Planning technicians require a minimum of four GCSEs or S Grades to enter the profession. Geography is an appropriate subject, and mathematics is a requirement.

Top Tips for getting into planning

- Get yourself aware of local planning issues. Most households get free newspapers put through the door these days. Look at the issues covered, and pay particular attention to the role that the planning office plays.
- Choose a real focus for a local planning issue in your GCSE or A-level geography project. Contact local interest groups. Talk to the planning office. Through a project of your own, you will begin to appreciate the complex role of the planner.
- Organize a period of work experience at a local planning office, perhaps via your school or college.
- Check that the planning course that you choose is accredited by the RTPI.
- Pass your driving test! Given the number of visits and meetings in various locations that planners have, being able to drive is a definite asset.

Case Study

Sue read geography at university because it was her favourite subject. After graduation, she had not decided what she would do for a career, so she completed a number of temporary jobs, like telephone research and secretarial work. Meanwhile, she had seen a job advertised in a national newspaper for an executive officer in the Food Section of the Ministry of Agriculture, Fisheries and Food. She got the job, and started about six months after graduation. Sue spent an enjoyable 18 months with MAFF but had decided by now that she really wanted to work in planning. She made enquiries with a local district council. Although an advertisement eventually appeared in the local paper, the council contacted her as well. Sue got the job and was a planning officer there for four years. She then moved on to her present post in another borough as principal planning officer.

Sue is certain that geography helped her prepare for planning, and specifies teamwork, cartographic and graphic skills, computing and report writing, although she concedes that some of these skills could have been developed in another chosen discipline. She completed a postgraduate Diploma in Urban Planning on day release at a local university over a three-year period, and then achieved her professional chartered status as a member of the Royal Town Planning Institute.

Case Study

Simon took a gap year after A-levels, and for much of the year worked in his borough council's planning department. He learnt a lot from the experience, which convinced him he wanted a career in town planning and property management. He read geography at a university in the north of England, and once he had obtained a good degree he approached a firm of London surveyors for a job. Working in planning during his gap year was a real asset, and Simon was successful in his application. After four years as a town planning surveyor, he moved on to another job in the City as a planning associate partner via an advertisement in the Estates Gazette. After three further years, another advertisement in the same trade journal recruited him to work on a four-year contract as a development manager in Dubai, although he admits disappointment that his geographical studies never actually taught him where Dubai was! Simon also studied part time for a postgraduate MSc in property development and management, and he obtained professional status as an associate of the Royal Institution of Chartered Surveyors whilst in his first job.

Useful addresses and information

The Royal Town Planning Institute
26 Portland Place
London W1N 4BE
Tel: (020) 7636 9107
E-mail: online@rtpi.org.uk
Web site: www.rtpi.org.uk

The Institute publish an excellent glossy brochure, *Careers in Town Planning*, which includes some varied case studies. Their *General Information about Membership* lists accredited planning courses.

The Society of Planning Technicians
7 Kingsway
Ferndown
Dorset BH22 9QN

5 Cartography

Map skills are fundamental to geography. Consequently, all school pupils should have used an Ordnance Survey sheet in a geography lesson. They may not have been able to fold it back up properly but they should have used it! Maps have long been the tools of geographers – but not their exclusive preserve. However, users often fail to appreciate how much thought has to go into what area is required on a map, what scale is appropriate, how selective a representation is required, and so on. These are some of the easiest issues facing the cartographer.

Cartography, as defined by the International Cartographic Association, is 'the discipline dealing with the conception, production, dissemination and study of maps'. Traditional map-making is centuries old but an explosion of technology in the later part of the 20th century has revolutionized the scope of the imagery. Satellites and remote sensing have given us new perspectives of the globe, and the storage of huge amounts of data on computer has enabled an entirely new branch of the discipline to emerge: Geographical Information Systems. GIS are used to store, process and display information on a computer. GIS packages, composed of a database, statistical and mathematical analysing capacity and graphic display facilities, are now widely in use at the workplace. They allow a level of analysis not possible until recently. For instance, a local authority can store a record of the whereabouts of gas pipes, electricity cables and water pipes on a street map. Each distribution is separately stored, and can be displayed individually or in conjunction with any of the others.

The structure of cartography

The main distinction in employer for cartographers is between the government (the public sector) and companies (the private sector). The privatization of what were once branches of government does cloud this distinction. There are less than 3,500 government cartographers and no one knows how many working for private companies. There may be as many as 10,000. The greater use of technology in this area of employment means a contraction of opportunities, and most advertised jobs attract stiff competition.

The obvious government employer is the largest producer of paper maps, the Ordnance Survey. Beginning as an essentially military organization, it is today an executive agency, and the national mapping agency for Great Britain; the Ordnance Survey of Northern Ireland is a separate agency. The Ordnance Survey employs about 14,000 staff at its headquarters in Southampton with another 50, mostly land surveyors, located around the country in 80 small field offices. Other government employers are the Hydrographic Office, the Meteorological Office, the Department of Environment, Transport and the Regions, the Scottish and Welsh offices, Military Survey, the Ministry of Agriculture, Food and Fisheries, the Forestry Commission, the Natural Environment Research Council at the British Geological Survey and the Institute of Oceanographic Sciences, together with county, regional and district councils. The armed forces also require specialist cartographers to prepare maps and charts and to interpret intelligence.

There are a number of large national organizations in the private sector who require cartographers: British Telecom, British Gas, British Coal, the Civil Aviation Authority, the Central Electricity Generating Board and the Post Office, to name but a few. There are also 25 or so commercial publishing houses and map printers who produce wall maps, charts, road atlases and maps. There are a dozen land, sea and air survey companies. Planning consultants need cartographers, as do civil engineering contractors and the exploration departments of oil companies. The service organizations like the AA and RAC require cartographers. A small number of cartographers are employed in the universities.

The jobs

The cartographic profession is subdivided into editorial and production work. Editorial work involves the verification, interpretation and application of the sources available before any map production can take place. Editors have to manipulate the background data to suit the needs of the user. Once they are satisfied that the level is appropriate, the source is transmitted to production personnel.

Production work has traditionally been the preserve of the draughtsperson. Traditional manual skills are still important but are now supplemented by use of GIS technology. The emergence of this technology has definitely opened up the commercial sector to cartography, and job advertisements can often be for a GIS manager rather than for a cartographer.

An example of a cartographic job is a mapping and charting officer, employed as a civil servant in one of the departments or agencies. As well as the obvious maps and charts, he or she creates graphics, technical drawings and illustrations, lecture aids, display materials and complete publications.

Allied professions

Planner, surveyor, information and computer technologist.

Personal qualities

Cartographic work is generally desk-bound and workstation-based. You need a genuine feeling for maps and a dedication to research. Good powers of concentration, patience, precision and attention to detail are necessary. Ideally, you need artistic flair and a sense of design. Because of the growing influence of technology, a knowledge of computer systems and commercial packages helps considerably. Indeed, a flexibility to working with traditional and also new methods is particularly appropriate.

Getting started

Editorial work in cartography usually employs graduates; production work by draughtspersons does not.

The most direct route into cartography for a geographer is by way of a specialist degree. There are not many to choose from – only one university offers a degree in cartography, two offer topographic science, three offer mapping science and three more GIS. Geography is an appropriate A-level in each case. If you take a geography degree, it will include an element of map usage, but 10 university courses offer a substantial amount of cartography. For someone wishing to go into this job area at the end of his or her geography degree, a diploma or MSc in cartography will be an asset.

BTec HND courses in GIS have emerged in the last few years, and also provide a job route for would-be cartographers.

The number of full-time diploma courses at technician level is an obvious source of qualification for draughtspersons, and consequently in-service training has been reduced. Two-year full-time BTec/SQA National Diploma courses are widely offered at colleges, covering cartography, surveying and topographic studies.

The civil service recruits non-graduates. Requirements for mapping and charting are two GCSEs/S Grades (A–C/1–3) for technicians grade 2, and three GCSEs/S Grades (A–C/1–3) or a BTec National Award for technicians grade 1. NVQs/SVQs may also be accepted. Officer trainees require five GCSEs/S Grades and two A-levels/three H Grades. Mapping and charting graduate trainees need a degree in a relevant subject: land surveying, computer science or geography.

Top Tips for getting into cartography

◆ Experiment with maps yourself. You may enjoy browsing through atlases or drawing maps in lessons but imagine creating images from scratch. Where would the data come from? How selective would you be?

◆ Get to know an atlas. Think of the demands on the cartographer in representing the required data.

◆ Find out what you can about Geographical Information Systems. How do they work? Who uses them, and why?

> ◆ Try to arrange work experience in a cartographic section of a government department or private company. Your local authority or water/electricity/gas company are your most likely sources.

Case Study

Jason *chose to read geography at university as it was 'the only subject that really interested me'. While doing his degree, he became fascinated by satellite images of the earth and hence he decided to do an MSc in Geographical Information Systems when he graduated. He was aware that geography was becoming increasingly computerized and recognized a growth area where he could combine his interests with a future career. As part of his Masters degree, he persuaded a major software company to fund his thesis on the introduction of image processing software into Europe. When Jason had completed his course, the company gave him a temporary contract to work for them for a year. Meanwhile, the authors of the software set up a regional office to the south-west of London. Jason went to work for them, and is now managing director of the office, supervising a team of 20 people.*

Case Study

Matt *left school with three A-levels, including geography – but he had no wish to go to university. He saw a job advertised in the local paper for an assistant post at the Mapping and Charting Establishment, got it and worked there for two years. He then moved to central London to work for the Department of the Environment, where he became a mapping and charting officer. He had realized the benefit of a degree for his career progress and started studying part time for a degree in geography. Eventually, he moved into the private sector as a GIS analyst. The company that he works for provide commercial GIS packages to insurance companies on all aspects of natural hazard risk assessment. 'The 1997/8 winter storms are a good example of the work we do. We interpreted Met Office weather data and provided the insurance companies with an impact study which they used to consider adjusting their premiums.' He finds that his role as a senior analyst 'requires an understanding of geographical theories and processes, especially in terms of spatial analysis techniques and natural hazard processes'. Matt is an enthusiast for geography, returning* *each year to his old school to represent the subject together with GIS at a careers convention, and he is currently completing a PhD in physical geography.*

Useful addresses and information

The British Cartographic Society
Department of Cartography
Oxford Brookes University
Gipsy Lane
Headington
Oxford OX3 0BP

The society publishes a very comprehensive pamphlet called *Careers in Cartography*, priced at £1. It contains details on the nature of cartography, of cartography courses and of prospective employers.

Military Survey
Elmwood Avenue
Feltham
Middlesex TW13 7AH
Tel: (020) 8818 2192

6 Environment

Widespread care and concern for the environment are relatively recent trends in society. Hence, many of the environment jobs now available are new, but unfortunately they are insufficient to satisfy the interest shown amongst young people. European Union policies are becoming increasingly sympathetic to the environment, and these are having an influence in the UK. The government, for instance, is showing a more enlightened attitude to agriculture. Meanwhile, the UK leads the way in accepting the resolutions made at Rio in 1992 on sustainability, biodiversity and a reduction in emissions into the atmosphere. In addition, Agenda 21 has provided a framework for local government to organize people to play a role in improving their environment.

Much environment-related employment is technical, hence specialist knowledge is frequently scientific. However, geographers tend to have a good, balanced view of things, which makes them particularly suited to environmental management.

The structure of the sector

Numerically, the environmental sector is dominated by volunteer workers, of whom there are 100,000 in conservation, compared to 12,000 full-time staff. There are three areas of employment – environmental protection, waste management and nature conservation. The main employers in the first two are government agencies, local authorities and industry, while nature conservation is dominated by statutory bodies and the so-called non-governmental organizations (NGOs), like the National Trust.

We can subdivide the sector by potential employer.

Government and statutory sector

An increasing responsibility for environment is reflected through the agencies that regulate its protection. Much of the work is multi-disciplinary, experts working with colleagues who have different expertise. At the national level, responsibilities lie with a number of bodies:

◆ The Department of Environment, Transport and the Regions (DETR) is the major department of state responsible for policies affecting quality of life, and hence the body has a central role for setting the policy framework for the environment in this country. It is also a funding provider.

◆ English Nature advise the government on nature conservation and protect the natural environment from development by promoting the conservation of natural habitats and landscape features. They manage nature reserves and identify and notify Sites of Special Scientific Interest.

◆ The Countryside Agency (formerly the Countryside Commission) promotes an understanding and enjoyment of the countryside of England through 300 staff at its headquarters in Cheltenham and in seven regional offices. The Commission also provides grant aid to support the salaries of over 1,000 countryside conservation and recreation staff.

◆ The Countryside Council for Wales, Scottish Natural Heritage and the Environment and Heritage Service in Northern Ireland have the combined powers of English Nature and the Countryside Agency in their respective parts of the United Kingdom.

◆ The Environment Agency is a regulatory authority for air and water, and hence protects and manages the environment of England and Wales. The Scottish Environmental Protection Agency does similar work in Scotland. Rivers, pollution and waste regulation are particular responsibilities.

Other government agencies are outlined below:

◆ The Forestry Commission is Britain's biggest estate manager. Its 1,750-person workforce is responsible for over 2 million acres of land in England, Wales and Scotland. The Commission advises ministers on policy and is responsible for implementing that policy. It gives grants to encourage tree planting. Forest Enterprise is a specific division of the Forestry Commission, and is responsible for the management of the nation's forests and woodlands. The Forest Research Agency provides high quality scientific research and surveys, and promotes high standards of sustainable forest management.

◆ The Farming and Rural Conservation Agency has been formed to assist government in the design, development and implementation of policies on the integration of farming and conservation, environmental protection and the rural economy.

◆ The Natural History Museum, the Royal Botanical gardens at Kew and Edinburgh, and the National Botanic Garden of Wales provide more specialist employment.

At the local level, planning, environment and leisure and recreation departments of local authorities need labour. Some local authorities manage country parks. National Parks are also administered within the local government system, although it is central government that provides most of the financing. On a day-to-day basis, they are run by National Park authorities.

Jobs in local government are not expanding, however. Many of their projects are put out for tender to private companies. With this increased privatization generally, much research work may be contracted out to universities or private consultants. This gives the agencies more of an administrative role. The Natural Environment Research Council (NERC) is the government's key environmental research organization and includes a number of smaller research institutes. Field survey work is carried out in order to advise government how to develop policy and legislation. Landowners and the general public are also advised.

Business and industry

Some private companies, particularly the large ones, employ in-house ecologists and environmental managers. Likely sectors in which to find these companies are mineral extraction, timber, energy, water, waste disposal, civil engineering, leisure and tourism.

Environmental consultancy provides a range of services on a contract basis to organizations not employing specialist staff. There are now several hundred consultancies, both large and small, in the UK.

Non-governmental organizations (NGOs)

These are the many well-known organizations that make up the voluntary sector. They have paid staff who need a great deal of help, often on a part-time, seasonal and volunteer basis. They rely on membership fees, donations and fundraising to support what they do. There are four categories of organization: wildlife bodies owning or managing land; campaign and pressure groups; practical conservation organizations; and the larger research organizations. Examples abound:

Wildlife
The National Trust
The National Trust for Scotland
Royal Society for the Protection of Birds (RSPB)
Wildfowl and Wetlands Trust
Wildlife trusts, which exist usually at county level

Pressure groups
Council for the Protection of Rural England (CPRE)
Friends of the Earth
Greenpeace
The Marine Conservation Society
The Ramblers
World Wide Fund for Nature

Practical conservation
British Trust for Conservation Volunteers
Conservation Volunteers for Northern Ireland
Scottish Conservation Projects Trust
Groundwork trusts

Research
Botanical Society of the British Isles
British Trust for Ornithology
Birdlife International

The jobs

Environmental job roles may be divided broadly, though not entirely satisfactorily, into administrative, technical/scientific and practical conservation. Governmental roles are likely to be the most administrative, with a lot more time spent in the office than in the field! Environmental matters are much higher-profile now than ever before. An increasing amount of public relations expertise is employed at governmental level, whilst administration assistants produce reports, publications and leaflets for public consumption. Environmental management in the office tends to be supported by a similar technical base to that of planning, namely through library, cartography and Geographical Information Systems. Pure research may be limited at the governmental level to what goes on in the research institutes. Government scientists and advisers may be called upon to give evidence at planning inquiries.

English Nature illustrates the range of job roles in a national organization. It employs policy specialists, whose brief is national and who specialize in areas like transport, tourism, geology and biology. At the local level, the team carry out practical land management roles at National Nature Reserves. It also employs conservation officers who carry out a multitude of roles, from working with landowners to carrying out nationally agreed initiatives or strategies at local level.

It is in the private sector that technical specialism is most used. The larger companies employ people with a science background. Ecologists, for instance, may be involved early on in large civil engineering projects. The supervision of plans to restore degraded land to maximize nature conservation value is another role. Implementing accredited environmental standards in industry is yet another. More controversially, the monitoring of genetically modified crops is a good example of specialist scientific monitoring. Water is an example of an industry in the private sector with recent increases in environmental opportunities. The regional water companies employ people to manage their own recreation sites, water pollution control officers and river corridor surveyors.

Rather than employ specialists, private companies may use environmental consultants on a contract basis. They may carry out field surveys on basic distributions of flora and fauna, for instance. Environmental impact assessments are frequently required these days to report on the consequences to the environment of a potential development. Consultancies can provide the necessary expertise to carry out such studies, and their more senior staff may give legal and financial advice of a specialist environmental nature.

However, it is for employment in practical conservation that the sector is most attractive and this takes us directly into the environment. Local authorities employ specialist teams who are probably part of the planning department. Likely job roles include rights of way officer, or definitive map officer, to designate and maintain public footpaths and bridleways; tree, forestry or arboricultural officer to manage the council's responsibility for its trees; and ecological or environmental officer to map habitats, collect data, assist conservation and provide interpretation for the public. More forward-looking authorities may employ a waste management officer, a recycling officer, a sustainable development officer or a local Agenda 21 officer.

National agencies, local councils and the voluntary sector all employ a range of practical employees to manage designated areas. Overall responsibilities here lie with countryside managers, wildlife managers, nature conservation officers, project officers or area officers, whose main role is development and forward planning whilst ensuring that the main remit of the organization is carried out locally. They also respond to consultations from outside bodies, from agriculture to planning departments. On a day-to-day basis, of course, they are responsible for their wardens (in reserves) or rangers (in parks), and their estate workers. They are responsible for the state of the countryside and its visitor service. This brings them into contact with the public a great deal. They are assisted by estate workers who carry out the manual labour, like woodland management, clearing scrub, and construction and maintenance of fences, car parks and footpaths.

The protection and management of countryside for the public has led the associated bodies and organizations into the provision of resources for education and interpretation. It is common now for some sort of visitor centre to be constructed on site, and for there to be an education officer in residence. An information and interpretation officer may be employed to promote a site. There may be some other expertise within this sort of employment. For instance, National Parks employ their own ecologists for advice and to undertake research, surveys and monitoring.

The pressure groups employ campaigners and parliamentary lobbyists. These roles actively involve formulating an organization's policy and producing briefings for politicians and the media. Such vital strategic roles usually require people with media or public relations experience.

Throughout this chapter, the assumption has been that jobs are found in the countryside in the UK, but there are local projects with a distinctly urban setting. The Groundwork Foundation in fact focuses on environmental regeneration in our towns and cities. There are also a number of well-advertised environmental projects abroad, set in threatened environments like rain forests. The more accessible ones are run by organizations or charities based in the UK. They essentially rely upon volunteers in the field but they also need qualified expertise to oversee the projects, as well as administrative assistants in the UK. Volunteering for environmental work abroad is included in programmes offered by some gap year organizations.

Those with aspirations on the environmental side should ensure they volunteer to work with one of the non-governmental organizations early on, to get valuable experience. Such groups have become high profile and there is much demand to work for them in a paid capacity, so volunteering provides an unpaid first rung on the experience ladder. Many of the groups will have local or regional offices near you, and indeed there may be a volunteer bureau in your town to give you advice. The most likely organizations to provide volunteering are indicated in the 'Useful addresses and information' section of the chapter. An additional advantage of volunteering is leaving a lasting impression of the quality of your work. Also, you might just be lucky enough to be on the spot when a vacancy arises.

Allied professions

Forester, farmer, landscape architect, town planner, estate manager, land agent, research scientist, ecologist, local authority leisure services and parks department officer.

Personal qualities

A good understanding of environment and of environmental principles is fundamental. A real commitment to the job is imperative. You need good communication skills, tact and diplomacy to deal with the public and to liaise with landowners, bodies and sometimes conflicting interest groups. You will need to work on your own initiative on occasions. Strength and fitness will come in very handy, as will a driving licence. A belief in environmentalism and in the purpose of the project on which you are currently working will at times need to sustain you through long, hard days in the field.

Getting started

There is no standard route into environmental management and conservation, given the tremendous diversity of expertise required. It is therefore logical that a science degree will give you most scope for employment. Geography does provide that broad perspective once again, and is a particularly suitable background to administrative and managerial posts. The Countryside Agency, for instance, names the subject as one of those it considers appropriate for would-be employees.

For a geographer with science A-levels, the choice of an environmental degree is an appropriate one if you have firm ideas about what you want to do. In choosing, look carefully at both the course title and description. Environmental sciences will stipulate one or two science A-levels, especially in the universities demanding higher grades. Environmental studies and management tends to be less specific, and a few places actually stipulate geography as a required A-level. Completing a postgraduate environmental qualification will enhance your chances.

The alternative route to A-levels is the Advanced GNVQ in land and environment, preferably combined with a science A-level. Higher National Diploma courses tend to focus upon particular aspects of the environment, like resource management or countryside recreation. A Modern Apprenticeship in environmental conservation is available, leading to at least NVQ level 3, and a framework for a National Traineeship in the environment is under development. This will offer training leading to NVQ level 2. There are a number of BTec conservation and countryside management courses at colleges of agriculture. Vocationally, the National Trust offers a careership training provision in countryside management (nine places currently) leading to NVQ at levels 2 and 3 in environmental conservation.

Top Tips for getting an environmental job

♦ There are many well-qualified people looking for a job in the environment. Voluntary, unpaid work is an obvious first step – not only for career tasting but to begin to build up relevant experience.

♦ Keep a look out for short training courses offered by some of the organizations. They will enable you to gain real practical skills like footpath management.

♦ Don't forget that some practical countryside experience is offered on conservation holidays, and even on projects abroad.

♦ If considering a career in environment, science A-levels are most appropriate choices – **if** you are capable in those subjects.

♦ Look carefully at the title and the description of environmental courses. They can vary considerably.

♦ Jobs tend to be advertised in the local press, Job Centres, the *New Scientist,* national newspapers (*The Guardian* on Wednesdays and *The Independent*) or specialist publications (the *Environment Post* and the *Countryside Jobs Service* weekly listing may be in your local library).

Case Study

Natasha says she loved geography from a young age. With encouragement from her parents to 'do what you enjoy because it will lead to a career you enjoy', she was unhesitant about pursuing the subject. She also did business studies and art with art history at A-level, and then went on to study the subject at a university on the south coast of England.

She advises that 'for students wishing to go into an environmental career, geography provides an excellent starting point in its breadth. It is a fascinating subject and it provides good all-round employees. However, to gain employment it usually needs to be backed up with enthusiasm, personal interests, voluntary work, and I would recommend doing scientific subjects with it at A-level. English and biology would also be a useful combination, followed by an MSc in environmental sciences or a more specialist course in remote sensing or marine resource management or countryside management.'

Natasha did some shop work immediately after graduation and then spent four months in Ukraine with Teaching Abroad, an organization advertised at the university. She stayed on in Ukraine to complete some independent environmental research in Kiev and Odessa, which was sponsored by groups in England. This experience helped her to get into environmental consultancy on her return, as an environmental scientist with consulting engineers based in East Anglia. After two years, she was relocated as the first environmental scientist to be based in the firm's 'out office' in Devon, and was seconded for two days a week to the Environment Agency South West Region as part of an environmental assessment team. She worked for the company for nearly four years.

Natasha is currently employed as Teign Estuary officer. The local district council formed a partnership with the Environment Agency to prepare a management plan for this Devon river estuary, and that partnership has now been extended to include five separate bodies. While based with the district council, she has an independent role that reflects the need for balance in creating the plan. This is where her geography comes in again.

'Geographers should have a holistic viewpoint, balancing environmental, social and economic understanding. I am dealing with people from different backgrounds – scientists, managers and business people like fishermen and harbour authorities. Geography does develop in you a sense of understanding of their different perspectives, and how to bring them together,' she maintains.

Attempting to put her geographical background into context with her career experiences, Natasha concludes: 'Things like project preparation – issue identification, report writing and dealing with questionnaires – developed through my geography, but I have also learnt to implement projects and plans and to manage them successfully through working. Practical experience is very important. At present, it all seems to come *together in what I see as that broad perspective again: a fascination for natural processes and the way people interact with them!'*

Case Study

*After doing English, biology and geography at A-level, **Kate** went on to do a teacher training degree at a college in the north-west of England. However, after two years on that course, she swapped to a BA degree in geography. She was keen to work in the environment, so after completing her degree she did an MSc in rural resource management at university in north Wales. Since then, Kate has worked exclusively in environmental jobs. Firstly, she was a ranger for Forest Enterprise in South Wales for nine months, a job she saw advertised in the* Guardian. *Since then, she has had two jobs with Exmoor National Park Authority. The first was as a ranger, and there were 560 applicants for the job! After two years in the job, Kate became a farm liaison officer. The post was advertised internally, and she has done it for the last three years.*

Of her present job, she says: 'I now specialize in working with the agricultural sector to promote conservation. The National Park Authority operates several grant schemes to encourage practical conservation work and I am in charge of administering these, which involves giving advice as well as bringing other National Park staff such as the archaeologist or the ecologist, to advise on more specialized subjects. As well as working directly with farmers, I also liaise with the National Farmers Union, the Ministry of Agriculture, Fisheries and Food and the Farming and Rural Conservation Agency, and other conservation bodies such as the wildlife trusts and English Nature. More recently, I have become more involved in rural development, as we are concerned at the fragmentation of the rural community due to the crisis in upland farming. If we have no farmers in the hills, there will be no one to continue the management of the countryside that we have finally begun to work on so well together!'

What of her geography? 'My career and the reason that I chose geography are linked in that I am very much an outdoor person and always wanted to do a job involving working in the field. Geography gave me an excellent foundation to understanding the forces that have produced the environment that we know today. It has definitely helped me to understand the complex interrelationship between the physical landscape and the human communities. This is essential for working in a protected area.'

What does Kate think about geography and an environmental career? 'The environmental sector is so diverse and complex that it is difficult to link it with any degree subject. Having studied geography, it may be necessary to do a more vocational postgraduate course.'

Useful addresses and information

The British Ecological Society
26 Blades Court
Deodar Road
Putney
London SW15 2NU
Tel: (020) 8871 9797
Fax: (020) 8871 9779
Web site: www.demon.co.uk/bes

The Institute of Ecology and Environmental Management
45 Southgate Street
Winchester SO23 9EH
Tel: (01962) 868626
Fax/answer: (01962) 868625
E-mail: enquiries@ieem.demon.co.uk

The above two organizations jointly publish *Rooting for a Career in Ecology and Environmental Management*, available from either of them. Please send a self-addressed A4 envelope with stamps to the value of 45 pence.

*The British Trust for Conservation Volunteers
36 St Mary's Street
Wallingford
Oxfordshire OX70 0EU
Tel: (01491) 839766
Fax: (01491) 839646
The British Trust for Conservation Volunteers will give you details of your local branch to advise on what projects are available nearby, and a volunteer pack, which includes a booklet of environmental training opportunities and details of working holidays.

The Council for the Protection of Rural England (CPRE)
Warwick House
25 Buckingham Palace Road
London SW1W 0PP
Tel: (020) 7976 6433

The Countryside Agency
John Dower House
Crescent Place
Cheltenham
Gloucestershire GL50 3RA
Tel: (01242) 521381
Fax: (01242) 584270
The Countryside Agency publish the free booklets *Employment and Training Opportunities in the Countryside* and *The Countryside Training Directory*.

The Countryside Commission for Wales
Plas Penrhos
Ffordd Penrhos
Bangor
Gwynedd LL57 2LQ
Tel: (01248) 385500

English Nature Enquiry Service
Northminster House
Peterborough
Cambridgeshire PE1 1UA
Tel: (01733) 455100
Web site: www.english-nature.org

The Environment Agency
They produce an informative booklet *Recruitment Information*. Ring their general enquiry line 0645 333111 for your nearest regional office.

*The Groundwork Foundation
85–87 Cornwall Street
Birmingham B33 3BY
Tel: (0121) 236 8565
The Groundwork Foundation works through a network of local schemes especially in urban areas, and they will put you in contact with your nearest one.

*The National Trust
36 Queen Anne's Gate
London SW1H 9AS
The National Trust publishes a leaflet *Working with the National Trust*. Send a sae not less than 4½ by 6 inches to The National Trust Membership Department, PO Box 39, Bromley, Kent BR1 3XL. Details of its training provision is available from The Careership Office, Lanhydrock, Bodmin, Cornwall PL30 4DE (tel: 01208 74281).

*The Ramblers Association
1–5 Wandsworth Road
London SW8 2XX
Tel: (020) 7582 6878

*The Royal Society for the Protection of Birds
The Lodge
Sandy
Bedfordshire SG19 2DL
Tel: (01767) 680551
Fax: (01767) 681284
The Royal Society for the Protection of Birds publishes *Careers in Conservation* for £5.

*The Woodland Trust
Autum Park
Dysart Road
Grantham
Lincolnshire NG31 6LL
Tel: (01476) 74297

WWF UK (World Wide Fund For Nature)
Panda House
Weyside Park
Catteshall Lane
Godalming
Surrey GU7 1XR

* organizations likely to offer environmental volunteer work

Further reading

Shepherd, A (1998) *Careers Working Outdoors,* Kogan Page, London
Working in Environmental Services (1997) COIC
Working in the Voluntary Sector (1999) COIC
Green Volunteers: The world guide to voluntary work in nature conservation (1998) Vacation Work Publications

7 Travel and tourism

In a recent advert for a business travel consultant, candidates were required to have amongst other skills, 'a good geographical knowledge'. You rarely see such stipulations in advertisements. However, it provides a reminder that in the old-fashioned sense, geography is all about places.

The travel industry has become one of Britain's biggest growth industries, as our average income and the amount of leisure time have increased. It is also a huge industry globally. The sector should be, in short, just the job for geographers.

The structure of the industry

This is one of those sectors that is difficult to delimit. Employees of airlines, coach and rail companies provide a service within travel and tourism but they are not included here. Broadly, travel covers the shop-window, selling side of the industry. Tourism covers companies putting together tours.

If we follow the natural sequence for a paying visitor, the industry is primarily composed of travel agency, tour operator and tourist board. Travel agencies are essentially retailers selling the products of airlines and tour operators. There are 7,500 ABTA (Association of British Travel Agents) registered outlets, each employing half a dozen people. There are a few large operators, providing more of a management structure.

Tour operators arrange travel, which is usually sold as a package, either directly or indirectly via a travel agent. Representatives do sometimes accompany some trips, and couriers and representatives are employed abroad. In fact, nearly 3,000 reps and couriers are employed by British travel firms, yet they are the visible minority of their company. They are outnumbered by UK-based administrative and clerical staff.

Tourist boards promote the area they represent. The UK operates internationally, attempting to sell the UK as a tourist destination. The four countries of the United Kingdom have their own tourist boards. Administratively, they monitor standards of accommodation and tourist attractions. Collected statistics help them to predict future trends. Tourist information centres promote their local facilities to the public from a variety of workplaces: council offices, libraries, museums and railway or bus stations. They offer more employment: over 2,000 regular staff work in the 800 tourist information centres throughout the UK. In addition, there are many seasonal staff. There are more tourism jobs in some areas than others, with coastal resorts and scenically beautiful regions standing out.

Local authorities also employ people to develop the tourism of a smaller area. They will either be employed on small self-contained projects or be part of a larger department like leisure services, planning or heritage.

The jobs

Travel consultants and travel agency clerks spend much of their day at the counter, discussing holiday ideas with members of the public. Besides booking holidays for them, they advise on health requirements, visas, insurance and foreign currency. Most of the business is centred on package holidays. Work at the sharp end of the business is office-based, and can be frenetic at times. Some travel agencies specialize in business travel, and this is usually conducted on the telephone.

Tour operators are most likely to be represented by couriers or tour managers and resort representatives. Couriers and managers travel with holidaymakers, and ensure that things run smoothly. They do the routine administration, dealing with tickets, seats and bookings, as well as providing commentaries on places of interest en route. As transfer couriers, they may meet in-coming flights and take passengers to check them into their hotels, returning them to the airport at the end of their stay. Resort reps usually spend a whole season at one location. They welcome parties, give details about the resort and the various trips and excursions that can be organized. They need to be on hand to deal with problems, and usually organize entertainment.

One of the more recent developments in tourism is adventure travel. Consequently, specialist travel companies have been set up or expanded to satisfy demand. They require sales people or travel consultants with some specialist knowledge or experience of independent travel or of particular parts of the world. Gap year organizations have also expanded, and the larger of these are now developing specialist roles for training, recruitment and marketing. Ecotourism, a more ecological approach to holidays, is also a recent phenomenon, which requires the occasional employment of specialist knowledge.

Tourist information centre assistants deal with enquiries and give out information to the public about local places of interest and how to get out and about. They liaise with local guesthouses and hotels to provide accommodation. Much of the employment is seasonal, geared to tourist demand. The assistants are employed either by the tourist board or the local authority. Special tourist guides may be required, although they are often freelance in status.

Some local authorities have a special team or part of an existing department to develop their area's tourist potential and to boost the local economy and employment. A tourist development officer, tourism and leisure officer or tourism marketing manager heads such a team. There are only a few such jobs, but jobs as assistants may be available. Today, heritage forms the basis of some tourism so even the most unlikely of industrial regions may employ tourist staff!

There are always a few jobs within a sector that do not conform to the broad categorization. Travel journalists and writers do work in the industry, for writing about places publicizes them and the outcome is to encourage more travel to those destinations. Journalists work for national newspapers and specialist magazines. Individuals also compile travel guides, for which they need a knowledge of a particular region. The journalists have probably followed a more general writing career, while the travel writers come from a variety of origins. They do not conform to an accepted career path, and they are rarely geographers. Nick Crane is the only well-known travel writer who is a geographer by training!

Allied professions

Hotel worker, foreign exchange clerk, air cabin crew member.

Personal qualities

Much travel and tourism work involves meeting the public, and hence entrants need to be friendly and polite, and able to work under pressure. Those who travel with the public need to be adaptable, well organized and reliable. More specialist tour companies may require specialist knowledge from you. A second language, preferably that of the country of foreign travel, is an asset, as is a good geographical knowledge.

Getting started

This is a sector where much full-time work does not require degree entry. Many companies recruit 16-year-olds, who work their way up the career ladder. The travel geography option at GNVQ level would be a good choice of qualification. The Travel Training Company provide a training programme within the industry. A mix of training and experience can lead to NVQ/SVQ levels 2 and 3 in travel services.

Most entrants come in at junior level, beginning as counter clerks or equivalent. A working knowledge of travel geography is an asset, and the most useful GCSE passes are English, mathematics and a foreign language. A-level entrants may come in at the same level. There are few specific graduate routes, although a few larger companies like American Express and Thomsons do have graduate training schemes.

BTec travel and tourism is perhaps a more specialized route; The SQA National Certificate has modules related to tourism. HNC and HNDs and the SQA Higher National Awards offer courses in tourism and travel, as well as business and finance with a tourist option. Appropriate degree courses include tourism and leisure management but recreation management and leisure studies are also relevant.

Tourist guides employed by regional tourist boards enter after the age of 23.

Many people do not see working in this sector as a long-term prospect, particularly those working abroad.

Top Tips for getting into travel and tourism

- ◆ Next time you go on holiday, try to see the organization of things from the representative's point of view. Would you like to be doing his or her job? If so, look into the sector in more detail.
- ◆ Holiday play schemes, or similar, are the best way to 'test yourself out' in dealing with the day-to-day organization of an itinerary with customers.
- ◆ The practical side of tourism is particularly suited to younger, independent individuals. If you are keen to get into tourism, do so early on in your career!
- ◆ If you are interested in a career in tourism, a foreign language is always a potential asset.
- ◆ Many tourist companies are very small, so career progression can best be achieved by 'job hopping'. Design your own career path.
- ◆ A hobby might just provide the basis for starting you off – bird watching or scuba diving, for example. Operators always want someone who understands clients' needs.

Case Study

Jo studied geography, business studies and classical civilization for her A-levels at school, and she went on to take geography at a university on the south coast of England. She took a gap year after university and travelled to Australia, an experience she says considerably broadened her outlook on life.

Since then she has worked in the travel and tourism sector, firstly as a reservations agent for a firm booking worldwide car hire and subsequently for a large business travel company in the south of England as a human resources assistant. She obtained that job through an employment agency.

Jo is a recent graduate, and so she feels that she is still not settled on a career. Of her geography, she has definite opinions however. 'Geography can be a very social subject with a massive variety of options available to suit everybody. For me these subjects have related to everyday life and experiences. If you feel there are certain aspects of geography that you do not excel in, this is fine as there are many courses that you can tailor-make for yourself. I chose not to select physical options. A geography degree prevents you from pigeon-holing yourself into a certain career. It does not limit you in any way when it comes to selecting a career path in the future. I studied human geography and I feel that people do not recognize or realize what this subject included. I studied, amongst other things, criminology, politics, architecture, transport policy, feminism and anthropology. These subjects have considerably heightened my social awareness, as many are very topical in everyday society. I thoroughly enjoyed my course. Each subject was so different and so topical.'

Case Study

Nicki took geography, economics and German at school. With good grades, she went on to read her favourite subject geography, at a university in the west of London. She began by studying a general geography degree but found the physical material too scientific, realizing that she had reached her limits at A-level. She changed her degree to human geography and was able to concentrate upon the sort of topics she preferred. She mixed more traditional elements with ones that were new to her, like the philosophy of geography.

After her degree, Nicki feels that she came out of university fairly unprepared for employment. 'The careers advice was there but I did not seek it out, and if I had my time again I would do more to enhance my employability. When you are fighting for graduate jobs, you realize that you are not so special after all!'

Nicki took a gap year after university, travelling and working in Australia, New Zealand and parts of the Far East. That experience, she feels, built up her self-confidence and leadership skills, and introduced her to other cultures and values. 'I suppose geography inspired me to look at the broader picture in life. It certainly opened my eyes to the wider world, which then encouraged me to travel,' she explains.

When Nicki returned to the UK, she obtained a job in travel sales in London, working for a specialist provider of long-haul packages. Nicki wanted to move into the marketing sector of travel, but to have stayed with the same company in order to achieve it would have meant moving out of London, something that she didn't want to do. She therefore made a sideways move into information technology. Although she learnt an awful lot about IT in 18 months, she felt that it was not the long-term career for her.

Nicki has recently moved back into the travel sector, to one of the development charities that provide gap year placements, in a newly created post of marketing co-ordinator. She is still sorting out her role but knows that, in marketing, she has found the sector that she really wants to work in. She will be helping to develop a marketing strategy, and she is only too aware that the breadth of commercial experience she has already gained in sales and IT, and in office environments generally, will help her in her new role.

Useful addresses

British Tourist Authority and the English Tourist Board
Thames Tower
Blacks Road
London W6 9EL
Tel: (020) 8563 3219
Send a stamped addressed envelope for careers information.

Wales Tourist Board
2 Fitzallan Road
Cardiff CF2 1UY
Tel: (01222) 499909

Northern Ireland Tourist Board
59 North Street
Belfast BT1 1NB
Tel: (01232) 231221
Web site: www.ni-tourism.com

Scottish Tourist Board
23 Ravelston Terrace
Edinburgh EH4 3EU
Tel: (0131) 332 2433
Web site: www.holiday.scotland.net

The Institute of Travel and Tourism
113 Victoria Street
St Albans AL1 3TJ
Tel: (01727) 854395
A careers information pack costs £3.

The Travel Training Company
The Cornerstone
The Broadway
Woking
Surrey GU21 5AR
Tel: (01483) 727321
Please enclose a stamped addressed envelope, if you send for
information.

Further reading

Collins, Verité Reily (1996) *Careers in the Travel Industry*, Kogan Page,
    London
Collins, Verité Reily (1999) *Getting into Tourism*, Trotman
Sharon, Donna and Sommers, Jo Anne (1997) *Great Careers for People
    Interested in Travel and Tourism*, Kogan Page, London
Working in Tourism and Leisure (1998) COIC

8 Development

Gone are the days when geography provided you with the opportunity to study regions of the world in detail. It is perhaps the topic of 'development' that comes closest today, through subdividing the world rather simplistically into more economically developed and less economically developed countries. The rich world-poor world division has complex economic, social and political causes, and its possible solutions always provoke debate in the classroom. It also stirs in some pupils a wish to do something about world poverty.

Let it be said at the outset that this emotion in itself is not enough for working in development. Simply wishing to help eradicate poverty offers little of practical use to those agencies who employ people in the field or, indeed, to the world's poor. However, what is needed more than anything is people with a skill or a field of expertise to offer. This is particularly true of work in development as opposed to relief. Development concerns long-term community-based projects, while relief projects are short-term, complex emergencies. Roles to play in the latter are short-lived and insecure.

The structure of the sector

The easiest way of categorizing the availability of jobs is by scale of employer. There is an international scale of activity, which is dominated by the United Nations and the World Bank. They do recruit, but usually on a senior level of expertise. The UN, for instance, administers its own development programme and also participates in a joint programme with a number of specialized agencies. The

UN is in financial crisis, and will not be recruiting significantly in the near future.

At the national level, the UK government has a major multilateral commitment to development, and the Department for International Development (DFID) is the focus. This is now a ministry in its own right, evidence of the government's interest in the sector. Indeed, the DFID is committed to the internationally agreed target of halving the proportion of people living in extreme poverty by 2015. More generally, employment in international relations at a governmental level is provided by the Foreign and Commonwealth Office, which looks after British interests abroad. There are about 4,000 jobs based at home, and about half that number in the Diplomatic Service, working in the country's embassies around the world. As with the DFID, all government ministries now recruit separately. The British Council, basically the country's cultural agency overseas with offices in over 100 countries, is another source of employment.

More recently, there has been a growth in private sector consultancy in technical, economic and social spheres. The nature of chosen projects will determine the area of work. A civil engineering company could be hired by an African government to carry out an impact assessment on a planned project, for instance. Of course, this reinforces the more general careers advice that employment in specific sectors can take you to parts of the less developed world through your work.

However, it is the non-governmental organizations (NGOs) that are the most obvious groups working in development. These are charities like Oxfam and Christian Aid, organizations that combine increasing awareness with fundraising. Their administrative work is predominantly in this country and involves the co-ordination of short research projects, fundraising, campaign work, community development and environmental work. The majority of the workforce is employed in this country, supporting projects run by a few in various parts of the world.

The jobs

At the international level of development employment, jobs are available but they tend to be at the higher and expert end of the

market. The United Nations does occasionally recruit expert advisers in industrial and economic development, social services and natural resources. The World Bank recruits people with several years' professional experience. The Bank does, however, also recruit some young professionals on its graduate programme. Such multilateral work is likely to be administrative and office-bound, in New York for the UN or Washington for the World Bank.

At the national level, two or three years' experience and a professional qualification or a degree are required to contribute to the Department for International Development's work. Assignments of two or three years are typical, and the most common sectors of work are agriculture, architecture, education, engineering, finance, fisheries, forestry, health and population, management, social development and surveying. There is also a graduate entry programme (see the 'Getting started' section later in the chapter). The British Council does not run general recruitment campaigns for school leavers or graduates as such. It does not have a graduate training programme, nor does it offer work experience placements. Vacancies do appear in the national press from time to time.

In education, the most direct involvement with North-South issues is by teaching and research in a university, contributing to development courses in undergraduate and postgraduate degrees. University lecturers will develop their own study areas of somewhere in the less economically developed world as part of ongoing academic research, while they are sometimes contracted to carry out consultancy work related to their region of expertise.

Teaching abroad offers accessible involvement in a key sector of any poor country. The British Council does employ teachers abroad on fixed-term contracts, and independent voluntary societies send out experienced teachers to areas of the developing world. Voluntary Service Overseas is a well-known example. It is often argued that education is the key to the future of poor countries, and teaching is a very practical way of contributing. The lack of resources available for use in the classroom can be staggering, but education is valued. So too will be your contribution.

The key functions of a charity organization in this country are fundraising, lobbying, research, media and publicity and financial administration, and a large number of professionals will be required to do these tasks. The philosophy of a charity may be philanthropic but the work culture is distinctly an entrepreneurial one. A small number of education or information officers exist in the development charities to provide educational information on the topics and countries covered by their charity, particularly to schools and youth groups. These posts are often filled by geographers.

The relief charities do, of course, employ people on the ground. The British Red Cross, for instance, employ 120 in 39 countries worldwide. Field officers are frequently organizers of successful projects, a skill that has come to be recognized more recently as logistics. More specifically, a project support manager will have overall responsibility, and logisticians will run it on a day-to-day basis. Aid workers work in refugee camps, and usually arrive early on the scene in many environmental emergencies (referred to as 'rapid onset emergencies') or political tragedies ('complex emergencies'). Indeed, it is a high-risk job, and it has become noticeably more dangerous in the last 10 years as civilians have become legitimate targets in conflicts. A list of organizations that employ volunteers overseas is provided at the end of the chapter.

Allied professions

Accountant, charity worker, fundraiser, logistician, social worker, teacher, youth or community worker.

Personal qualities

Whatever the job role in this sector, you need good communication skills for use at home and abroad. In the field, you need to be able to analyse situations swiftly, understand the political and cultural context in which you are working and cope with the stressful situations in which you may find yourself.

Getting started

Geography is a great springboard for working in development. If you were not convinced while at school, a degree in the subject and specialist options in the second and third years may well have changed your thinking. Although it may not be stipulated on job advertisements, a more specific postgraduate Masters course will complement the generalism of geography, and could, for instance, be in agricultural economics, demography or water engineering. Conversely, graduates of another discipline to geography may wish to 'convert' to the field via a Masters in development studies of some sort. As has been implied by what has already been said in the chapter, most posts in the field in this sector will go to graduates.

Most civil service recruitment is now carried out by the departments or agencies themselves, so applicants apply directly to DFID or the Foreign and Commonwealth Office. The exception to this is the Fast Stream Development Programme for people of high ability with the potential to progress quickly. A geographer with a good degree might use this route to get into the Diplomatic Service, for instance.

One of the real problems in getting a job in development is how to acquire relevant experience before you get that job. There are schemes that do help a graduate to get some experience. DFID run the Associate Professional Officer Scheme, which combines doing a Masters degree with practical training, usually including overseas experience. The scheme recruits annually but, frustratingly, overseas experience is asked for! The Overseas Development Institute offers 18 to 20 fellowships annually for economists or similar to work for two years in the public sector of a developing country. In addition, the World Bank and the Organization for Economic Cooperation and Development run Young Professionals Programs. The European Commission does recruit graduates as temporary trainees, called Stagiaires, to work for three to five months. This may help in the obtaining of a permanent post later on.

While it is certainly true that most jobs require graduates, voluntary sector jobs are many and varied. Hence, some roles will not require a degree. The agencies are particularly concerned to select the most suitable candidates for roles overseas, and they will provide professional training themselves. The British Red Cross, for instance, runs a prestigious week-long basic training course and trains the best applicants it singles out in international humanitarian law.

So, how do you obtain experience before your first job in development? Sensible first steps into development employment are holiday office-work experience on the administrative side, and volunteering on schemes in your local area for practical experience. After all, it should not be the location that is of paramount importance but how much use you can be! The aid agencies themselves recommend taking part in a gap year scheme. There has been a recent flourishing in the number of schemes, and you are able to choose widely from a number of destinations. Most of the schemes involve teaching for six months or so. For instance, in 1998–99 the Gap Organization sent 40 per cent of their successful applicants to various countries to teach English and 28 per cent to teach a variety of subjects in schools. Other gap schemes involve conservation, farming and social work. Sources for finding out more about gap organizations that provide placements abroad are given at the end of the chapter.

Top Tips for getting into development

◆ The Catch 22 of getting into development is that you need experience to get a job, so how do you get your first job? Get some first-hand experience of the developing world by travelling in long vacations, and consider the many gap year organizations offering participation in work schemes abroad.

◆ To gauge your suitability for the sort of aid work done abroad, volunteer for schemes in this country. Also, contact the aid agencies for office-based work experience. This can be done as part of a school scheme. It is never too early to get such experience.

◆ Apply to smaller organizations as well as the larger ones. They are less well known, and hence get fewer applicants.

◆ Be prepared in a practical sense: pass your driving test and have a first-aid qualification.

◆ Read *The Guardian* every Wednesday for jobs. Get to know the job market by reading the job adverts.

◆ Get to know the Government's current thinking on issues and specific countries by reading the policy briefings on DFID's Web site.

_____ **Case Study** _____

Ruth is a development education officer with the non-governmental organization Intermediate Technology, which has its headquarters in the Midlands. After studying geography, biology and English at school, she chose geography for her degree as she felt that it would be both challenging as a subject and useful and relevant to a wide range of careers. By the time she had completed her degree at Cambridge, Ruth had become particularly interested in development issues. She did so well in her degree that she was able to obtain Economic and Social Research Council funding to stay on to do a Masters course in development studies. Deciding to broaden her experience, Ruth then completed a Post Graduate Certificate of Education to enable her to teach. She spent two years in the classroom, teaching geography at a public school. She enjoyed the experience but felt that she wanted a wider challenge from a job.

Ruth applied successfully for her present post after she saw an advertisement in The Guardian. *'I enjoy my job so much because it is part of the public affairs arm of the organization and as such I feel able to contribute directly to their voice,' she claims. Her first year in the job has been a busy one. Her organization was asked to make an input into the National Curriculum review, a task that underlines the liaison role in her job. She also has a project management brief, the focus of which has been the development of an interactive Web site. Thus far, she has represented the organization at international conferences, and hopes to visit some of their offices in the developing world. Ruth feels that this is a good time to be in her job, as government agencies are building support to improve awareness of 'development' within the education sector.*

She feels indebted to the subject of geography for helping her to interpret and analyse information, and above all 'approach multi-faceted subjects in a holistic manner'. Ruth has continued to find these over-viewing skills useful in her project management in particular.

_____ **Case Study** _____

*After he had completed a geography degree at a northern university, **Dominic** was keen to follow a career in the development sector. He had studied two sciences previously at A-level, and decided to do an MSc in land and water management to give himself a more specialist skill. He then obtained immediate employment as a water technician on a scheme in Guatemala. Despite a valuable experience that confirmed to Dominic that he had chosen the right sector, the project placement was only for one year. It took him the same amount of time to get another job, so he had to be persistent. Finally, he became a co-financing assistant with*

Christian Aid. He is now a programme funding officer with them. Dominic has travelled widely in the developing world, and he particularly remembers setting up a BBC 9 o'clock News report on El Niño from Peru. His most recent work has been in Kosovo.

Dominic feels that geography has given him the broad background knowledge to appraise and monitor overseas aid programmes in the agriculture, water sanitation and health sectors. Although he is enthusiastic about the subject for a first degree, he thinks a second language is a real asset and that it is appropriate to get additional training to go into a profession, as he did. 'Get educated, and then trained,' he advises.

Case Study

Murray *did geography at university, when he became particularly interested in hydrology. Consequently, he went on to do an MSc at another university in water resources technology and management. He then completed three months of industrial experience with the Institute of Hydrology, followed by a job as an assistant hydrologist with the National Rivers Authority, now the Environment Agency.*

Murray wanted to work in the developing world and read about the Busoga Trust, an organization that funds the building of wells and installing of pumps to improve the quality of life in southern Uganda. He went out as a volunteer first, and liked it so much that he resigned his job and returned to Uganda. He is now head of the water and sanitation sections, based in the town of Jinja.

Murray feels strongly that development workers should not go out with the idea that they are the experts, and wishing to impose their ideas on people. 'The local water technicians have a wealth of knowledge of the work and of the needs of the rural population, and this is invaluable and deserves respecting,' says Murray.

Useful addresses and information

Department for International Development
Personnel Department
Room V240
94 Victoria Street
London SW1E 5JL
Tel: (020) 7917 0275

Foreign and Commonwealth Office
Recruitment Section, Personnel Management Department
1 Palace Street
London SW1 5HE
Tel: (020) 7238 4265/6/7/8/9
E-mail: pmd.fco@gtnet.gov.uk
Fast stream entry scheme details and entry forms are available from
Capitas RAS, Innovation Court, New Street, Basingstoke, Hants
RG21 7JB (tel: 01256 383683).

These organizations send volunteers overseas:

Skillshare Africa
3 Belvoir Street
Leicester LE1 6SL
Tel: (0116) 254 1862

Voluntary Service Overseas
317 Putney Bridge Road
London SW15 2PN
Tel: (020) 8780 2266

UNA International Service
Suite 3a
Hunter House
57 Goodramgate
York YO1 2LS
Tel: (01904) 647799

International Cooperation for Development
Unit 3, Canonbury Yard
190a New North Road
London N1
Tel: (020) 7354 0883

Further reading

A number of gap year organizations have schemes in various parts of
the less developed world. The following books may be useful:

Doe, T and Evans, H (1998) *A Year Off... A Year On?* Lifetime Careers
Butcher, V (1997) *Taking a Year Off,* Trotman

9 Land, water and atmosphere

The scope of this chapter seems pretty comprehensive. Jobs on land, in water and in the atmosphere, where else is there to work? What the chapter actually covers is job ideas related in some way to the physical geography you will have come across in the classroom. So, if you are interested in rivers, coasts, glaciation, soils, slopes, volcanoes, earthquakes or meteorology, read on. From the outset, let's try and put physical geography into some sort of context. In order to understand the workings of the physical environment, it is most appropriate to combine geography with the sciences. You will probably be aware that in studying particular physical topics, geography combines well with individual sciences: for soils and ecosystems, with biology; for volcanoes and earthquakes, with physics and chemistry. Hence, when we examine entry to many of the professions mentioned in this chapter, the science qualifications may well be even more dominant than geography.

So, what relative importance do geography and the sciences have in getting into these careers? Think of it in terms of geographers keeping their feet firmly on the ground. The further from the ground a career goes, the less geographical the specific requirement. Geomorphology and surveying are core geography. However, go deep under the surface and you are moving through the geosciences into the realms of applied chemistry and physics. Equally, rise above the surface and you come to meteorology. Despite all that coverage of weather in geography, when it comes to working in the field, employment through a geography degree alone is rare. An understanding of the physics of the atmosphere and the use of complex mathematical modelling determine the main academic subject requirements.

This chapter deals with a tremendous variety of careers, which, for simplicity, will be dealt with separately. Nevertheless, there will be areas of employment in the real world where several disciplines work together in a team. We will begin on the earth's surface, and move gradually away from the core of geography.

Geomorphology

This is the study of the shape of the Earth. A-level geography will probably have encompassed the term. A study of the physical processes that have created that shape will most likely have been exemplified by hydrology, coastal processes and glaciation.

The structure of the sector and the jobs

Jobs in geomorphology exist most obviously in the universities, where lecturers combine teaching commitment with applied research. The latter may well lead to consultancy work in either the public or the private sector. A good example of such work is advising on the slope stability of a hillside that a new motorway route is to cross. Taking such a career path stems from the options you choose in the second and third years of your geography degree. The safest professional advice is to complete a Masters degree in the specialist field of interest, as this will give you a measure of professional competence.

Given that Britain does not experience active ice processes, jobs in glacial geomorphology are limited. The British Antarctic Survey, based in Cambridge, is perhaps an exception. BAS carries out research in Antarctica of global significance to earth, atmospheric and life sciences in its three permanently staffed stations. It employs geographers, as well as physicists, chemists, biologists and geologists. Coasts are actively managed, and in places the rapidity of coastal processes is of concern. There is some role for the coastal geomorphologist, mainly in consultancy contracts but also in employment for consultancy companies, local authorities that administer coastlines and even MAFF, the Ministry of Agriculture, Fisheries and Food, which is responsible for national coastal strategy.

The greatest scope for employment is in hydrology. The obvious focus for an academic interest is the Institute of Hydrology, run by the Natural Environment Research Council. The Environment Agency employs specialist hydrologists and hydrogeologists in flood defence and water resource management. The latter function illustrates the agency's statutory role, balancing water abstraction by water companies with protecting vulnerable rivers and springs. Hydrologists analyse field data, and measure river flow, groundwater levels, rainfall and evaporation. Mathematical modelling is used to problem-solve and to predict future demand.

Commercially, the biggest employers are the water companies. Water is a very large job sector. In England and Wales, there are 17 supply companies employing over 5,000 people, and 10 service companies employing 33,000. In Scotland, three water authorities employ 7,000 staff, and the Department of the Environment Northern Ireland has over 2,000 employees. Add to that the 10,000 working for the contracting companies and you have some idea why the water industry is the most likely employer.

Getting started

For a geography graduate, the most likely water profession is hydrologist. The obvious route in is via a water company graduate training scheme. This applies similarly for other graduates who wish to enter specialist professions, for instance chemists, geologists, biologists and the various kinds of engineer. As has already been discussed in this chapter, geography graduates should pursue their geomorphological interests to an MSc or even a PhD level in order to go into more specialist areas of the field.

Beyond the specialist professions, entry to the water industry is less likely to be with a degree. Plant operatives and craftspeople, like water plant attendants and control room operatives, may enter with few or no qualifications, as training is provided on the job. NVQs/SVQs can be studied, or day release or block release used to study for relevant BTec/SQA certificates. Modern Apprenticeships are available, and in Scotland these are part of the Skillseekers Initiative.

At technical and supervisory grades, for instance bylaws manager or safety adviser, entry can be by GCSEs/S Grades, and employees will receive BTec/SQA-based on-the-job training. For higher technical grades and managers, in areas like distribution or water services, many people are promoted internally, but NVQs/SVQs at levels 3 and 4, HNCs and HNDs, and even degrees are possible qualifications.

Surveying

The surveying profession is concerned with the management and development of property in all its aspects.

The structure of the sector and the jobs

There are a number of specific types of surveyor, the most directly geographical of which are land, hydrographic, and planning and development surveyors.

Land surveyors are the ones you are most likely to see around, using theodolites and ranging poles. They may well be supplying cartographers, civil engineers or planners with required information by means of very sophisticated technology like aerial photography, satellite surveying and automatic surveying equipment. Land surveyors may be employed in private practice, by consulting engineers, by large construction companies, by aerial survey companies, or by the government's surveying departments – the Ordnance Survey, the Ministry of Defence Mapping and Charting Establishment and the Directorate of Overseas Surveys. The DOS does send experienced surveyors abroad, and it is reckoned that at any one time half the land surveyors are working abroad.

Hydrographic surveyors map the sea bed, rivers, harbours and ports. They too use state-of-the-art equipment, including sonar and echo sounders. Although the increased exploration of the North Sea for oil has increased the number of such surveyors for the private sector, most hydrographic surveyors are in the Royal Navy and the United Kingdom Hydrographic Office, which produces admiralty charts for the Navy and for shipping companies all over the world. Nevertheless, there are still fewer than 200 people working in the sector.

Planning and development surveying involves all aspects of urban and rural planning. Surveyors work as part of a team, offering advice on economics and amenities, conservation and urban renewal schemes, aiming to show what is possible with the resources and time available. They work particularly closely with planners, and are likely to be their work colleagues at district or county level. They may work for the Department of the Environment, Transport and the Regions, or for planning consultants, property developers or construction companies.

Other specialist surveyors are: **minerals surveyors**, advising on and valuing mining sites; **building surveyors**; **rural practice surveyors**, concerned with farms and estates especially; and **quantity surveyors**, calculating and controlling the costs of construction developments. Finally, there is **general practice surveying**, encompassing commercial property, property management, estate agency, valuation and chattels.

Getting started

Surveying is essentially a graduate profession. Most obviously, a degree in surveying is appropriate. A common route is a degree such as geography, followed by a postgraduate conversion course at MSc or Diploma level, which can take anywhere between one and three years. Candidates must also complete a minimum two years' supervised professional training and an interview in order to obtain professional status, eg as a member of the Royal Institution of Chartered Surveyors. It is possible to start work as a trainee technical surveyor with a minimum of four GCSEs/SCEs at 16-plus, but it is more usual to start at 18-plus with those qualifications together with A-levels/H Grades.

Landscape architecture

This is the most creative of the careers in this chapter, for the landscape architect is a designer of everyday space.

The structure of the sector and the jobs

This is a small sector, but growth is steady. There are around 1,800 fully qualified professional members. About half the profession work for government departments, public bodies, local authorities, industry and development corporations. The rest work for design and architectural practices, are self-employed or work for contractors or developers. There are good opportunities to work abroad.

A landscape architect will tender for work by producing design ideas for a client and an estimate of the costs of the project. If the landscape architect is successful in this bid, he or she will begin work on the project in earnest. The design will vary according to setting. In rural areas the focus is on agricultural, forest and tourist landscapes. Work in towns may involve the integration of areas of housing, road works, parks and regeneration. Indeed, the most obvious projects to work on are restored or regenerated townscapes where a significant space needs to be laid out from scratch.

Getting started

This is a graduate career. There are first degrees in landscape architecture but only at six universities, and several in landscape management. Again, geography is a common general degree, and probably needs to be followed by a postgraduate course. After that, graduates complete the Landscape Institute's professional practice examination (part 4) in order to become fully qualified professionally.

Civil engineering

Civil engineers are involved with the design, construction and maintenance of large structures. Traditionally, those structures have been unroofed but that distinction has broken down as some buildings have got larger.

The structure of the sector and the jobs

Civil engineers are now involved with any large construction project, which could be a transport system, a public health service project, structures like bridges, and flood and coastal protection. Civil engineers can be subdivided into branches of expertise that reflect: the nature of the project – water engineers or public health or highway and traffic engineers; the technical specialism – structural engineers (who deal with the stresses and strains of materials); or the employer – municipal engineers work for the local authority. The most directly geographical role of the civil engineer occurs early on in the job with site analysis – physical surveys of ground and soil, and more economic impact assessment.

Most civil engineers work for firms of consultants, construction companies or local or national government. Consultancies can vary in size from one self-employed person to a major employer of 2,000–3,000. Such firms are likely to be multi-disciplinary, although some big companies can be specialists.

Construction companies represent both the civil engineering industry and the building industry. The largest construction company in a project generally acts as the main contractor and subcontracts smaller specialist firms. Normally on a civil engineering project, planning, design and costing are carried out by consultants and the construction company builds. Both will be working for the client, either a large company or an organization. Local government employs engineers to plan, provide, maintain and manage infrastructure. The work may involve consulting the public, undertaking environmental assessments, producing designs, supervising the work, setting out contracts, negotiating finances and producing reports for consideration. Some engineers can work for national government, and in the armed forces.

Getting started

There are chartered engineers, incorporated engineers and engineering technicians. Chartered engineers have a degree and at least four years of experience. Civil engineering is the obvious degree discipline, although a variety of other subjects are accredited for status. Incorporated engineers have a degree, or an HNC or HND, and have worked at least three years in engineering. Engineering technicians provide technical skills in the industry, and qualifications are likely to be BTec/SQA National Certificate or Diploma, or Advanced GNVQ/GSVQ. Modern Apprenticeships may be available in some areas.

Geosciences

These involve the study of the structure, evolution and dynamics of the earth, and they are multi-disciplinary in their approach.

The structure of the sector and the jobs

Much of the work of geoscientists will have been popularized to you through plate tectonics and continental drift in geography. Since the 1950s, it has been a most exciting field of research to be in. Much of the study is, however, applied science. Geophysics determines the physical structure of the earth by using gravity, magnetism, seismic waves and radioactivity. Geochemistry examines how the elements behave when igneous and sedimentary rocks form. Geology itself is multi-disciplinary and very different from physical geography. The geosciences are likely to be employed in mineral exploitation, in engineering geology on dam, road or tunnel sites, in studies of natural water supplies, in waste disposal and in hazard study. Many of these are broadly geographical in scope but the knowledge and expertise are scientific.

Career opportunities are to be found in four main areas, and jobs can take you to all parts of the world. The first area is exploration and production – the search for natural resources such as fossil fuels, metals, construction materials and ground water, and the geological management of their extraction. Secondly, there are engineering and environmental jobs. The focus for these is the investigation and monitoring of local ground conditions associated with construction, planning, land utilization and environmental issues. Geological surveys are the third employment area, involving the systematic collection of surface and subsurface geological information, onshore and offshore, for the production of geological, geophysical and geochemical maps and databases. Finally, geoscience jobs are found in education and research – teaching and research posts in universities; geological and science teaching posts in schools and colleges; and museum posts. The biggest employers of geoscientists are the British Geological Survey, government departments, the water industry and teaching and lecturing.

Getting started

Again, most employment is by graduate entry, especially by relevant degree discipline. Geography can be studied to A-level, where it may be accepted for entry to some geology degrees. An HND can lead to employment at technician level. Within some universities good HND candidates can transfer up to the degree course in the final or the last two years.

Oceanography

This is the study of the seas and oceans, including shorelines, coastal waters, estuaries and continental shelves.

The structure of the sector and the jobs

There are few jobs in oceanography. Those that do exist are mainly in scientific research for the Natural Environment Research Council in two oceanographic institutions. The council also funds research units elsewhere. Other employers are MAFF, the Admiralty, the Royal Navy and the Scott Polar Research Institute. In the environmental sector, water companies and the Environment Agency are employers. There is some work for oil companies as environmental legislation is increasing.

Getting started

This is a graduate profession. Quite frankly, a geographer would need to begin to specialize with his or her first degree and/or to complete an appropriate postgraduate qualification. For, apart from ocean-ography, the other likely degree requirements for entry are scientific.

Meteorology

As referred to before, meteorology is a technical field popularized through the study of weather in geography lessons. Weather fore-casting is only one aspect, indeed specialism, of the science.

The structure of the sector and the jobs

Most employment is by the Meteorological Office, an executive agency within the Ministry of Defence since 1990. The Met Office supplies a wide range of weather information for UK military and civil use. The service is becoming increasingly commercial, as demand has broadened to require the supply of specific forecasts to a wide range of companies and organizations. The Met Office employs over 2,000 staff, half at its headquarters in Bracknell and the rest dispersed around over 80 different locations in the UK and a few abroad. Given an increasing dependence upon sophisticated computer systems, employment prospects are at best static. Other public sector employ-ment involves branches of the scientific civil service, employing meteorologists in oceanographic, hydrology and pollution institutes. The British Antarctic Survey, the Agricultural and Food Research Council and the Department for International Development provide opportunities for work abroad, and so do the UN technical aid programmes. Universities provide a few openings where fundamental research and teaching are carried out in an academic environment. The service industries are the main employers in the private sector, in water and energy supply especially. There are a few jobs in instrument manufacturing; and also a few in exploration, especially for the offshore oil industry, commercial airlines, and private fore-casting and consultancy firms.

Getting started

Graduate entry to the Met Office (22 out of 55 in 1995) is most likely via a degree in meteorology, mathematics, physics, computer science or electronics. Geography may be acceptable for an exceptional candidate who can show some ability in meteorology. Helen Young, one of the radio and television weather presenters, was a geography graduate who was recruited by the Met Office. The most obvious route in for a geographer is via a very good undergraduate thesis on the subject and/or an MSc in meteorology. The Met Office has its own college to ensure in-house continuous professional advancement. Job progression is from research to forecasting.

School leavers are recruited as assistant scientific officers, observing, coding and plotting the weather, and assisting forecasters. They must have a minimum of four GCSEs (A–C), but many also have A-levels.

Personal qualities

Although the types of employment are varied, they share scientific rigour and accuracy as requirements. Consequently, people with a professional attitude are sought. You need to communicate effectively, both orally and on paper. The ability to work in a team is common to all these jobs. You must be able to be self-sufficient when on your own, maybe in a harsh environment. Physical stamina is called upon in data collection, and computer literacy will be expected for analyses.

Top Tips for getting into physical geography jobs

- Take sciences at A-level with your geography.
- If you do a degree course in geography, start to specialize in physical geography options in your second and third years. Choose a relevant topic on which to complete your thesis or dissertation.
- Work experience and holiday employment are vital ways of showing a commitment to your sector of interest.
- If you are invited for interview, think back over the field-work you have carried out. Be able to tell the interviewer

---

what the purpose was, what you did, what results you obtained and what the work showed. Scientific methodology is a transferable skill.

◆ Keep a look-out for job advertisements in *The Guardian* and *New Scientist*.

Case Study

*Having studied geography, maths and physics for his A-levels, **Colin** went to university in the north of England to study geography. In the second and third years, he concentrated on geomorphological options. His dissertation was focused on a study of mass movement on the East Anglian coast.*

After university, Colin joined a high-street bank management training scheme. He worked for them for a year in a structured programme spread across the banking functions. However, he felt that banking was not for him and he looked for employment in the aspect of geography that had appealed to him most at university – coastal geomorphology.

In 1990, he joined the engineering section of the technical services unit of a district council in East Anglia. Colin was employed to undertake a wide variety of tasks in coastal protection, providing specialist information in geomorphology on coastal processes. Some of his work has been of a pure research type, including predicting rates of cliff erosion, and being a member of a study group looking at sediment movement in the southern North Sea. The main focus of his job, however, has been to help produce the council's coastal protection policy. More widely, he has contributed to engineering cost-benefit analyses of protection schemes, and has helped formulate responses to applications to extract gravels offshore.

Useful addresses and information

Informative careers booklets are available from:

The Civil Engineering Careers Service
1 Great George Street
Westminster
London SW1P 3AA
Tel: (020) 7665 2106/5

Fax: (020) 7233 0515
E-mail: careers@ice.org.uk
Web site: www.ice.org.uk
The service publish annually the factfile *Careers in Civil and Structural Engineering.*

The Construction Careers Forum
PO Box 976
London EC1V 1PB
The forum have available a free copy of *The Construction Industry Handbook of Professional and Management Careers.*

The Geological Society
Burlington House
Piccadilly
London W1V 0JU
Tel: (020) 7434 9944
The society has free booklets *Careers in Geoscience* and *Career Opportunities in Education for Geoscience Graduates.*

The Royal Meteorological Society
104 Oxford Road
Reading
Berkshire RG1 7LL
The society has free pamphlets *Careers in Meteorology* and *Meteorology in British Universities.*

The Royal Institution of Chartered Surveyors
Surveyor Court
Westwood Way
Coventry CV4 8JE

The Landscape Institute
6/7 Barnard Mews
London SW11 1QU
The institute have a booklet *Professional Careers in Landscape Architecture, Landscape Sciences and Landscape Management.*

The Board for Education and Training in the Water Industry
1 Queen Anne's Gate
London SW1H 9BT
Tel: (020) 7957 4524
Web site: www.betwi.demon.co.uk

The Meteorological Office
Job Advice Centre
Powell Duffryn House
London Road
Bracknell
Berks. RG12 2SX
Tel: (01344) 855243
Fax: (01344) 855260
A careers folder *Inside the Met Office* is available. The Web site www.met-office.gov.uk has information about the Met Office and all current recruitment campaigns.

The Ministry of Agriculture, Fisheries and Food
Management and Development Division
Personnel Management (Admin) Branch
Nobel House
17 Smith Square
London SW1P 3JR
Tel: (020) 7238 6383

Further reading

Working in Buildings and Property (1997) Careers and Occupational
    Information Centre (COIC)
Working in Construction (1998) COIC
Working in the Water Industry (1999) COIC

Transport

Transport systems could not be simpler in concept: taking passengers or goods from their origin to their destination. However, you will be aware of the complex problems involved in planning and running transport networks from geography lessons. The impact of motorway or bypass construction, or the sheer weight of traffic in city centres are good examples. The subtle balance between hard economics and social, political and environmental constraints is only the beginning of the complexities of a sector that requires long-term macro-planning and short term micro-solutions to everyday problems. In the present political climate, deregulation has resulted in more competition between commercial companies, less involvement from local authorities and fewer state-owned organizations.

The main modes of transport are convenient subdivisions of careers in this sector.

General transport administration

The structure of the sector and the jobs

In national government terms, transport is now administered by the Department of Environment, Transport and the Regions (DETR). The DETR has specific responsibility for the transport system in England, including cars, lorries, buses, roads, trains, ships and civil aviation. On some issues such as aviation or road safety, the department is responsible for policy for the UK generally. It has to ensure that transport is safe, efficient and environmentally acceptable, catering for the needs of industry, commerce and the general public. It also negotiates on Britain's behalf with other nations and within the European Union. From its headquarters in central London, staff also liaise with local authorities about developments in and funding of local transport. The department's executive agencies put aspects of policy into practice: building trunk roads, supervising vehicle certification, inspection and licensing, licensing operators and drivers, and monitoring marine safety.

Meanwhile, most local authorities have a highways department, which manages road systems and designs and builds roads. Traffic planners are employed especially at the regional level of government and sometimes in a partnership between local authorities. Some authorities also operate transport systems but these are separately managed.

Getting started

At national level, general civil service requirements are appropriate despite the separation of its component parts. Hence, the DETR will consider entry for administrative assistants with a minimum of two GCSEs/S Grades (A–C/1–3); and junior managers with two A-levels/ three H Grades and three GCSEs/S Grades (A–C/1–3). Higher executive and senior executive officers require a degree.

At local level, clerical and administrative staff may enter at 16 with four to five GCSEs/S Grades (A–C/1–3) for clerical posts, and at 18 with a minimum of two A-levels or a GNVQ/GSVQ advanced level award in business studies or public administration. For graduates, there are some formal graduate entry schemes. No specific discipline is required.

Air transport

Air traffic control

The structure of the sector and the jobs

Each of the sectors has a body that supervises it. The Civil Aviation Authority was set up to regulate safety, run air traffic services and license the UK air crew, air traffic controllers and maintenance engineers who are responsible for safety in the air, the airworthiness of craft and the development of airports.

The CAA and the Ministry of Defence jointly run the National Air Traffic Services (NATS), which control air traffic over Britain. These are crowded skies, as increasing numbers of us fly in and out of the country. It is the role of controllers to keep the traffic moving safely. The job is a complex pressurized one, which requires communication between controller and pilot. Electronic equipment does help, with radar displays showing an aircraft's position, track, height and number. Controllers need to work in three dimensions, with 'en route' services guiding pilots through a cube of air space and approach controllers enabling planes to land, often through a stacking system over a busy airport. Controllers work from two major centres at West Drayton, near Heathrow Airport, and Prestwick in Strathclyde. A new system will necessitate the relocation from West Drayton to Swanwick in Hampshire. Controllers are also employed at the London airports and the UK's regional airports. To be a controller, you need to be fit, have good eyesight, hearing and concentration, and be able to retain essential facts, remain calm and react rapidly to changing circumstances like the weather.

Getting started

The CAA recruits about 90 air traffic control officers annually. Assistants are recruited locally, and can retrain to become ATCOs, although there are age limits. NATS employ 1,700 people in air traffic control. CAA cadet entry is with five GCSEs/S Grades including English and maths, and two A-levels/three H Grades or a BTec/SQA advanced qualification. Entry age limits are 18–27. After a day-long programme of preliminary selection tests and finding out about the job, successful candidates are invited for interview. Candidates must pass a medical.

Airport management

The structure of the sector and the jobs

Airports are complex environments to manage. Consider the everyday components of an airport: passengers, airlines, air traffic control, immigration, freight handling, catering, car parking. Most traffic flows through the largest airports, so that is where the jobs are concentrated. The most significant owner and manager of airports is BAA (formerly the British Airports Authority) with seven, including the two largest ones at Heathrow and Gatwick, which handle 75 per cent of the UK's passenger traffic and 85 per cent of the freight market. BAA is now in fact a large commercial concern, which has diversified its operations into airport hotels, land and property development, and cargo and freight-forwarding. Each of its airports is run as a subsidiary company, so employees usually work at only one of them. The company has a small headquarters staff for forecasting and planning. The Civil Aviation Authority manages operations at several other airports, and local authorities look after 20 more, including Manchester, the third largest. The London City Airport is privately owned.

Airport managers have overall responsibility for reasonably small teams. Terminal officers assisted by traffic officers are responsible for terminal buildings. Traffic duty officers look after the interests of passengers and their luggage.

Getting started

Relevant qualifications and previous experience are the main recruitment factors. BAA at Heathrow take trainees in airport services, looking for GCSEs or equivalent in maths, English, CDT and science. GNVQs and HNCs are considered.

Individual airlines

The structure of the sector and the jobs

There are some 220 aircraft operators in Britain. Commercial management includes finance, marketing, personnel, management services and PR. Senior managers are responsible for strategy, route planning and developing charter work. Flight scheduling is a complex business.

Getting started

Promotion is mainly from within. Education to GCSE standard is required, with passes in maths, English and geography. A knowledge of another language is desirable, and keyboard skills are useful. Experience in a public contact job is essential. Graduates are recruited in small numbers. British Airways, for example, recruits for its business planning, engineering, finance, general management, pensions and purchasing sections.

Flying

The structure of the sector and the jobs

The most obvious sectors in which to fly are military and commercial.

The Royal Air Force has been cut in recent years to just over 50,000. Most British combat aircraft only require a pilot and a navigator. All pilots and navigators are officers, and need considerable skill and aptitude to fly and navigate modern combat aircraft. As well as 'strike' aircraft, there are transport, maritime reconnaissance, airborne early warning and helicopter operations. Only a small proportion of the RAF fly, and aircrew tend to spend little of their time in the air. Indeed, they tend to be most frequently in the air in their early career. Much pre-flight preparation takes place, simulators provide a cheaper way to train, and extensive post-flight analysis is usual.

There are about 30 commercial airlines operating in the UK. British Airways, one of the world's largest carriers, is the UK's major employer within commercial aviation. It has a fleet of 190 planes, including 7 Concordes. BA and partners employ over 50,000 people, including a quarter of the people holding a commercial pilot's licence in this country. Britannia is the world's biggest charter airline with 24 aircraft, and extra planes leased in peak periods. British Midland is the UK's second largest scheduled service airline, with a fleet of 35 aircraft. Virgin Atlantic is a more recent player.

Most planes are operated by a captain and a co-pilot, whose duties involve more time on the ground than in the air. Long-haul flights will involve stopovers, while up to 4 short-haul flights might be flown in a day. There could be other opportunities in commercial aviation, namely in the air taxi and charter business, flying helicopters or small executive jets, aerial survey work and photography, and crop spraying.

Getting started

Officer entry is between 17 and 29. Although age and qualification requirements vary between different branches, candidates must have at least two A-levels/Highers or equivalent, together with five GCSEs at grade A–C/1–3 to include maths and English language. About 75 per cent of candidates are graduates. Recruitment is a rigorous four days of aptitude tests, exercises, a medical and interviews. A good understanding of meteorology is required in training, and hence GCSE or A-level geography can prove a useful introduction.

In the commercial sector, precise requirements vary between airlines. BA require at least seven GCSE/S Grade passes including English language, maths and a science, plus two A-levels/three H Grades, ideally maths and physics. British Midland require five GCSEs/S Grades. Selection for all companies is rigorous. Many pilots have a degree.

Road transport

The structure of the sector and the jobs

The Highways Agency is responsible for managing, maintaining and improving the national motorway network and the trunk road system in England. It is an executive agency of the Department of the Environment, Transport and the Regions.

In road transport, the ratio of managerial, professional and supervisory staff to operative staff is high due to the many small companies in road haulage. Although transport journeys are dominated by the car, the road transport sector is a very fragmented industry. Bus and coach companies, and taxis and minicabs provide for passengers. Managers in public transport need to plan vehicle scheduling, service timetabling, crew rostering and so on. Road haulage dominates the transport of goods. Transport managers plan load, delivery patterns, routes, collections and runs. Computer systems have become indispensable.

Freight forwarders must have an expert knowledge of the transport business to find the best way of getting goods around the country and around the world. The planned result is a balance between time, value, weight, shape, size and perishability. They must be aware of current pricing, packing, custom and excise requirements, licences, regulations and insurance. Over 50 per cent of import-export firms use freight forwarders to improve efficiency and costs.

Logistics is a relatively new title, first used in the late 1980s, and is now commonly used in the road haulage business. It refers to the management of an entire supply chain, from raw materials to the point of consumption: purchase and supply, materials handling, materials management, production planning, production control, transport, storage, distribution, installation and servicing. It is a massive and expanding field. Logisticians work for manufacturers, retailers, specialist providers, consultancies and the armed forces. In the public sector, central and local government and the health service are ideal for such management. In the private sector, the food industry and retail chains benefit, as well as BT, British Steel, and the water and electricity companies. As the job title is so new, those working in it are frequently referred to as transport managers.

Getting started

The Highways Agency recruits along civil service lines as outlined earlier for the Department of Environment, Transport and the Regions. A high proportion of managers have worked their way up through company structures. There is direct administrative entry for school leavers at 18-plus via GCSEs/S Grades and A-levels/Highers, or BTec/SQA. Only the largest organizations recruit graduates, as in passenger services like London Transport.

There is no formal entry route into the freight sector. Most entrants start at the bottom, and need at least four GCSEs/S Grades, including English, maths and geography. A language is useful. There are A-level and graduate traineeships with larger firms. There are no well-defined career paths within logistics. Graduates are increasingly recruited. There are a few degrees in logistics, and several more offer modules as part of a general management or transport management degree. Hence at present, related disciplines like geography are welcome.

Allied professions

Armed forces employee, air steward, aircraft maintenance engineer, driver, warehouse manager.

Case Study

Greg *studied economics and English literature with his A-level geography at school. He had long wanted to be a pilot, and was faced with the choice that many have at that stage – whether to apply straight to an airline or to do a degree first for general flexibility. In fact, Greg applied to Britain's major airline but after several interviews he was rejected. He had meanwhile applied to university anyway, and went off to study geography in South Wales.*

He was determined to fly, and the company finally accepted him two years later. The university allowed him to suspend his degree and he entered the 18-month-long pilot's licence course in Scotland. Greg never completed his degree, and now flies long-haul 747s as a co-pilot.

Greg thinks back to his geography as 'a good scientific/methodological background whilst giving you wide options'. Not surprisingly, he has found the meteorological aspects of the subject most useful. 'The atmosphere is where I work, so its dynamics are crucial to my work. Whilst not perhaps influencing my getting the job in the first place, now I'm here it has been a very useful tool,' he concludes.

Case Study

Barry *read geography on the south coast of England. He always favoured physical geography by instinct, and it was his intention to work in a connected job in and around the university city where he had settled. Three months of unemployment after graduation made him less choosy however, and he saw a job in the paper with a US business information company. He got that job, and stayed for three years.*

Eventually, he felt that he had achieved all that he wanted to at the company. He had settled down in the city, and knew that he would have to move out of that employment sector for further opportunities. He saw another advert in the paper for a road cargo company that was setting up a new customer liaison section. He got that job, and ironically moved into a sector that has more obvious connections with geography. 'I think that the subject gave me a general spatial awareness, and this has been relevant to my job in the transport sector. I route vehicles through Europe, so an awareness of the European Community and its outline geography are very important.'

Barry has worked for the transport company for a number of years now, and he is only too aware that his A-level combination of geography, economics and maths was perfect for his job in what has become known as logistics. 'Only recently my company's name has been changed to incorporate that term!' admits Barry with a wry smile.

Useful addresses

Department of the Environment, Transport and the Regions
Recruitment Section
PSD
4th Floor, 4/01 Great Minster House
76 Marsham Street
London SW1P 4DR
Tel: (020) 7890 3225

The Highways Agency
Human Resources
Room 10/36
St Christopher House
Southwark Street
London SE1 0TE
Tel: (020) 7921 4367

The Institute of Logistics and Transport Supply-Chain Centre
PO Box 5787
Corby
Northants NN17 4XQ
Tel: (01536) 740100
Web site: www.iolt.org.uk

Further reading

Working in Airports (1997) COIC
Working in Transport and Distribution (1997) COIC

Working with people

Human geography courses usually begin with a study of population, as so much of the rest of the course is about the results of human behaviour on the landscape. In a consumer society most jobs will involve either working with other people or meeting the public, or both. Yet there are some jobs that focus upon people themselves. This can involve the analysis of populations, examining the behaviour of people as consumers, helping to give them favourable working conditions and catering for their basic requirements, such as housing.

Demography

This is the study of population, and is the starting point of a consideration of working with people.

The structure of the sector and jobs

The dominant employer in demography is government. The co-ordinating body is the Office for National Statistics, created in 1996 by the merger of the Central Statistical Office with the Office of Population Censuses and Surveys. It is responsible to the Chancellor of the Exchequer, and has a staff of over 3,000. It is the central administrator of the law on civil marriages and the registration of births, marriages and deaths. It also keeps central records, and controls the local network of registrars and superintendent registrars. Most familiar for geographers is the preparation of statistics on aspects of population totals, distribution and change, and elements like migration, fertility, births, marriages, deaths and diseases. The office is responsible for organizing the census every 10 years. The ONS also co-ordinates activity in the use of official statistics. The data collected are used more widely, particularly by other government departments that monitor existing policies and perhaps modify them on the strength of revealed trends. You can imagine that such data are particularly relevant to areas like education and health care. Indeed, the departments administering these areas also have research officers and statistical staff of their own completing research on population.

The Immigration and Nationality Directorate deals directly with people: the over 68 million passengers a year coming into the UK. Immigration officers assess the eligibility of these passengers to enter the UK. This involves the processing of immigrants' applications and dealing with appeals, work at airports and seaports, and checking passports and visas. New entrants are usually posted to Dover, Heathrow or Gatwick.

Local government may also use the findings of the census but there are no specialist departments in local government dealing exclusively with demographic trends. Administratively, the only real jobs at this level are those of registrars and superintendent registrars, supervising the registration of births, marriages and deaths.

The ethnic diversity of the UK population has led to the growth of various organizations representing the interests of minority groups. These exist at national level in a lobbying and co-ordinating role, and at local level as advisers and project workers. The Refugee Council is such an example.

There are a few jobs in higher education in demography, composed of the lecturers who research in the field as well as lecture in the universities. They sometimes complete research work on a consultancy basis for organizations that commission it.

Getting started

There are few demography degrees, so most graduates will have studied another subject. Given the usual inclusion of demography in a geography degree, the subject can provide the ideal broader context. The Office of National Statistics and the Immigration and Nationality Directorate recruit along civil service lines. For instance, an executive officer needs three GCSEs/S Grades (A–C/1–3) and two A-levels/three Highers, or equivalent.

Market research

The most obvious commercial application of an interest in population is market research. Finding out what the customer wants and why is the key to success in the consumer society. However, market research has a wider sphere than consumption, and is relevant also to industry and to social and political matters.

The structure of the sector and jobs

More than half of the jobs available in market research are in agencies. Most of the 300 or so agencies are in London, and they are dominated by consumer research. The agency makes a proposal to a potential client and, if successful, carries out a study on the required topic of interest. Background data are collected, interviews conducted and a report written up. Agencies can be made up of varying sizes of workforce, from a few to a few hundred. Many large industrial and commercial concerns have their own departments. Planning and organization in market research are carried out by executives, who will have started as research assistants. The bulk of project work is carried out by many interviewers, whose jobs are part time and irregular, and who are employed purely to carry out questionnaires on the public.

Social research takes place mainly in local and central government, in higher education and in institutes like Social and Community Research. Most government research is carried out by the Social Surveys Division of the Office for National Statistics.

Getting started

No particular academic qualifications are required, although many entrants have a degree. The most relevant degrees are in social sciences like geography, business studies, maths, statistics and economics. It is estimated that about 250 new vacancies occur every year for graduates. The Market Research Society's Foundation and Certificate examinations can be taken as part of a degree, HND or A-level/H Grade. They are equivalent to NVQ levels 2 and 3 respectively.

Human resources

Human resources or personnel management attempts to make the most of the company or organization's workforce.

The structure of the sector and the jobs

All organizations of any real size, ranging from commercial and industrial concerns to the leisure sector, have human resources officers. Personnel work involves the recruitment of new people, and training and keeping existing ones. It also involves ensuring that organizations deal with their employees consistently, fairly and within the terms of employment law.

Recruitment and selection involve drawing up job descriptions, preparing and checking application forms, interviewing, and selecting or rejecting candidates. Training and development provide training programmes for employees. Industrial and employee relations involve meeting trade union representatives to negotiate pay and conditions. Employee services like awareness of health and safety matters, maintaining staff records and personal counselling also fall within the scope of human resources.

Getting started

The majority of human resources officers have a degree, from a wide range of disciplines, or a Higher National Diploma. It is possible to obtain promotion with experience in a clerical or secretarial post in a human resource department. Relatively few organizations recruit graduates directly into personnel work. Rather, they tend to recruit them into general management training schemes, which include personnel experience.

Housing

The housing market has changed significantly over the last decade, as local authorities have lost responsibility for much of their housing stock. Former tenants have bought homes, and non-profit-making housing associations have emerged.

The structure of the sector and the jobs

Most jobs in housing exist with a local authority and the rest are with housing associations and trusts, and some private companies. As well as housing, local authorities maintain and improve thousands of properties, and they own shops, community halls and sheltered housing. They also have a statutory duty to house or provide temporary accommodation for certain groups of homeless. Housing associations have grown in number, and over 3,000 are registered. Local authorities and housing associations have similar roles to play.

Housing officers manage the stock of rented housing, and this covers leases, tenancy agreements, rent collection, repair work, and appointing and managing caretakers and wardens. Today, local authorities may have a more pivotal role in enabling other organizations to meet local housing needs. Hence, the job probably involves monitoring and assessing housing needs and outlining how to meet those needs. The balance between renovating old properties and building new ones is likely to be a concern for long-term planning, and managing needs in relation to restricted budgets is a short-term priority. Some housing officers specialize in areas of special need – homelessness or disabilities, for instance.

Getting started

Qualifications are frequently not specified. Many entrants have degrees or BTec/SQA Higher National Awards.

Allied professions

Social scientist, statistician, careers adviser, occupational psychologist, training officer, advertising employee, marketing employee, public relations employee, community worker, estate manager, social worker, surveyor.

Case Study

Susan is human resources director for a large publishing group. After her geography degree, she joined a multinational food group via the 'milk round'. After a number of roles in human resources in the company, she became European personnel manager. She has moved companies three times to achieve her present role.

'Geography has helped me to analyse and present complex data and to prioritize social and commercial factors. My studies also taught me to speed-read large quantities of reports and identify key points,' Susan maintains. She also found the population dynamics and cultural issues in the social geography she studied helpful when working on the Continent.

Susan concludes that 'Geography is all about balancing facts and issues concerned with how the world operates in its widest sense. It's about understanding these factors and arriving at sound conclusions and recommendations for a way forward. A graduate with these skills is well equipped for a role in the commercial world.'

Useful addresses

The Office for National Statistics
Personnel Section
4th Floor
1 Drummond Gate
London SW1V 2QQ
Tel: (020) 7233 9233

Immigration and Nationality Directorate
PMD – Recruitment Section
Room 804
Apollo House
36 Wellesley Road
Croydon CR9 3RR
Tel: (020) 8760 8242

Institute of Personnel and Development
IPD House
35 Camp Road
London SW19 4UX
Tel: (020) 8971 9000
Web site: www.ipd.co.uk

The Market Research Society
15 Northburgh Street
London EC14 0AH
Tel: (020) 7490 4911.

Further reading

Working in Local Government (1997) COIC

Working in the media

The information explosion is a leading characteristic of modern life, and the way that we think and act has come to be dominated by the media: the newspapers we read, the radio we listen to, the television we watch, the computers we use. This may not seem to be a sector so obviously linked with geography. However, the range of skills developed in the subject is very appropriate to working in the media. The everyday nature of the subject matter of geography is also relevant. A recent advertisement for a research assistant asked for a graduate in 'a subject with a strong emphasis on current affairs', to be which geography can make a perfectly legitimate claim.

The media jobs selected here are essentially creative. Radio and television also include production, technical, and management and administration elements not covered.

Journalism

The structure of the sector and the jobs

Journalists work in newspapers, magazines, broadcasting and press agencies. Events tend to determine the day for newspaper journalists. They visit places, interview people and research to gather information for a news story or 'copy'. Once typed, now largely by word processor, the story goes to the paper's sub-editor for editing and comment. Each day is geared to the deadline, which journalists have to work towards. The deadline is not so frequent for magazine journalists, who might be writing for a weekly or a monthly publication. They work for consumer publications or for business and trade magazines. Staff journalists write news reports and features, and the specialism of particular publications may necessitate employing specialist freelance writers as well. For a magazine, a sub-editor similarly edits, writes headlines and prepares the journal for printing. A press agency journalist supplies news stories, features and photographs to newspaper or magazine publishers.

Journalists also work for national and independent radio and television companies. Local radio tends to be a first step for many – working as a reporter, preparing scripts and interviewing people. Journalists tend now to be trained in 'bi-media techniques', appropriate to both radio and television.

Getting started

Entry to journalism is via a number of routes. There are direct-entry courses for those who gain employment with a newspaper and then receive training. Minimum requirements are five GCSEs/SCEs (A–C/1–3), including English or equivalent. Most applicants tend to have higher qualifications. There are full-time pre-entry courses, and these require two A-levels/H Grades or equivalent, including English at GCSE/SCE or A-level/H Grade. Over half of successful applicants these days are graduates.

Researchers

The structure of the sector and the jobs

Researchers work in radio (radio research assistants) and television (television researchers). They are employed to work on particular programmes, and it is the nature of the programme that determines the work involved. Geographers will find a current affairs or documentary programme particularly suitable. Researchers find suitable topics, people and events for the programme. They may well go out and film material themselves, including interviews, and also prepare scripts. For documentaries, they may carry out library research into a subject, which will be provided for the programme's producer, editor, director or scriptwriter. They may well have to find suitable locations for filming (what better illustration of using a 'sense of place' for a geographer?), view old photographs, films or newsreel, and choose music. On a more practical level, they advertise for suitable people, have meetings with programme makers and look after interviewees when they arrive at the studio. In radio there is more of an emphasis upon using books, newspapers and sound archives. Many researchers move on to do other things – writing or journalism, programme editing or starting their own production company.

Getting started

Most researchers have a degree, usually in the field of research in which they specialize. Many jobs in radio and television go to people already working in the company. This means that most entry-level jobs are very junior and low paid. When they are advertised, they often specify quite modest qualifications. Short-listed candidates, however, often exceed those minimum requirements.

Public relations

The structure of the sector and the jobs

Public relations is all about presenting a favourable understanding of an organization to the outside world, be it to consumers, businesses, the financial community, employees or the public at large. A very large number of companies and organizations use public relations, from well-known commercial and industrial concerns to religious organizations and charities. Public relations is a two-way process, and gathering information about how a company or organization is perceived and feeding that back are also important elements of the job.

Companies or organizations may employ their own specialist in-house departments. Job titles vary: press officer, publicity officer, information officer, and their assistants. Alternatively, PR consultancies provide a service for a number of clients. A public relations officer discusses needs, and plans a programme accordingly. Writing and editing press releases, reports, articles and speeches are key skills, and liaison with the media is increasingly seen as a vital component of the role. Special events like press launches are usually organized by the PR department.

Getting started

There are no specific general requirements, but in practice most entrants have a degree. Any degree discipline is acceptable. There are some BTec HNC and HND courses in business studies with a PR specialism. They require either one A-level/two H Grades with four GCSEs/S Grades (A–C/1–3) or a BTec/SQA National Award.

―――――――――――― **Case Study** ――――――――――――

A curate

Rob did his geography degree in the north-east of England and then completed a year as a parish assistant in the same region. He went on to three years of ordination training, and is now a serving curate in a London diocese. 'Geography has developed in me the ability to produce a systematic line of argument and thinking, to work collaboratively in problem solving and to look at issues on a variety of levels.'

He also feels that the content of his university course was helpful: 'Courses on regional development and urban regeneration greatly influenced my ministry in inner-city Newcastle. Courses on humanistic and perceptual geography have informed my later thinking and research on the spirituality of place.'

―――――――――――― **Case Study** ――――――――――――

Radio and television producer

Tim is north of England producer for BBC TV and Radio News. His radio journalism background was forged in the same northern city where he is still based. He entered the profession via a BBC local radio trainee reporter scheme, having completed a geography degree at Cambridge. 'More than anything geography has given me the ability to gather information from a variety of different sources and draw conclusions – from the information point of view, 101 different geographical subjects! Almost all of the courses I did have been useful at some time in news gathering, and little nuggets of information are invariably useful.'

Case Study

Film producer

Jon graduated with a geography degree from a Midlands university in 1996, and he worked in the same city for one of the largest independent television companies. He has recently moved south to work for the BBC.

He has helped to produce programmes for Channel 4, BBC2, ITV and Sky. He admits that 'Geography has been very useful in my chosen career because of its broad base and sociological analysis. This may have been because of the excellent department I went to letting me explore avenues I found interesting. However, speaking to other people in the media who studied the subject, it would appear to be a general rule.'

Jon says of his recent work, 'As I have moved into documentary film making, my degree has helped enormously. My recent documentary looked at housing needs in the city superimposed on to themes of racial tension, poverty, drug abuse, exclusion from school and demographic divisions.'

Case Study

Newspaper journalist

Andrew did a degree in geography in the late 1980s, after spending a summer on a European grant travelling and researching the problems of immigrant workers in France and Germany. He seized the chance to travel both within the UK and abroad as much as possible, and completed his university dissertation on the British government's inner-city policy. He obtained funding to spend a year on a fellowship at Harvard after his degree, and then a further grant to work for the New York City government for a year. The experience he gained in working abroad and in publishing *numerous freelance articles helped him obtain a place on the* Financial Times *graduate training programme. He has worked for the newspaper through the 1990s, as a companies writer, general reporter, accountancy correspondent, and latterly in the paper's Paris and Moscow bureaux.*

Andrew says, 'Studying geography helped to develop my powers of analysis and synthesis. Its current affairs and international relations content gave me a taste for life in different parts of the world, which has led directly to my being a foreign correspondent.'

Significantly, all the individuals in the sample went directly into their career after finishing their geography degree. They did not need to take a further qualification to enter. Hence, the subject was accepted as an appropriate one, and the strength of the individual was the key factor.

Further reading

Kent, Simon (1997) *Careers in Journalism*, Kogan Page, London
Short, Dean (1998) *So You Want To Be a Journalist?: An insider's guide to career opportunities*, Kogan Page, London
Working in Journalism (1998) COIC
Selby, Michael (1997) *Careers in Television and Radio*, Kogan Page, London
Hird, Caroline and Grigg, Joanna (1996) *Careers in Marketing, Advertising and Public Relations*, Kogan Page, London

13 Geography and the commercial sector

It would be misleading to focus this book solely upon jobs with a geographical content when the evidence clearly points to many geographers going into the commercial sector, where they display their acquired skills rather than their knowledge. Chapter 2 outlined the match between the skills desired by employers and those encouraged by studying geography.

It would be beyond the scope of this book to examine in detail the range of possible commercial jobs open to geographers, as many graduate jobs are non-specific of degree discipline. This chapter is somewhat different in format from the preceding chapters. A series of short case studies is used to show what areas of commercial life some geographers have moved into and what advantages they feel the subject has given them. The experiences are accompanied in most cases by summary details of the associated careers.

Although the subject matter of geography does not always have a connection with the commercial world where many geographers make their careers, it appears that the skills developed through studying geography are recognized. Such skills were outlined in Chapter 2.

Corporate finance

Case Study

Neil moved on from his geography degree to a career in finance. He was recruited by a major accountancy firm at the 'milk round', one of the careers fairs held at major universities at which large employers attend. Entering as a trainee, he rose to become a corporate finance manager. He was recently 'headhunted' by a broadcasting company. 'I developed the capacity to express myself well on paper through geography, and the empathy for different cultural perspectives it encourages certainly equipped me to work in a multinational company.'

Banking

Case Study

Tim is area customer services manager with a high-street bank in South Wales. He joined the bank on their graduate training programme straight from doing geography at university. 'Geography in the commercial world is more about skills than knowledge, and these have been the most critical benefits. The lateral thinking it has instilled in me is the most interesting. Problem solving, 'thinking outside the box', the research mentality and teamwork are all assets for geographers to bring to banking.'

Over the last 15 years, banking has diversified considerably into offering a wide range of financial services. No longer do you merely deposit money and receive interest, or take out a loan and pay interest. Banks now also commonly advise on investments, insurance, mortgages and personal and business finance. The high-street banks are the best known but other prospective employers are the Bank of England, investment or merchant banks and international banks.

Getting started

There are no minimum requirements as such for customer service assistants but many banks ask for two to four GCSEs/S Grades (A–C/1–3), normally including English and maths. It is possible to gain promotion to management, but most entrants realistically hope to reach a supervisory level. A-level entrants usually aspire to junior management posts and above, while graduate trainees target senior

management via the 'fast-track' route. A wide range of degree subjects is acceptable, including geography.

Accountancy

Case Study

*After a geography degree in Scotland, **James** did a series of temporary jobs and was unemployed for short spells of time before becoming a trainee accountant with one of the large accountancy companies. 'The use of computers is very useful in my present job; presentation skills are also used at work; the skill to write reports clearly and concisely was developed at university.'*

Accountants manage the financial affairs of individuals and organizations, and advise on matters of taxation, investments and record keeping. It is a numerate profession but it also involves communication with clients, and hence people skills are important. Financial accountants deal with accounting records and their analysis. The resulting reports refer to profit and loss accounts, budgets and cash flow forecasts. Financial accountants also oversee the payroll, internal audits and stock control. Management accountancy deals with analysing the running of a business and the production of a report. Accountants work for companies and for public services.

Getting started

Accountants qualify by passing examinations to join a professional institute. This can be the Institute of Chartered Accountants, the Association of Chartered Certified Accountants, the Association of International Accountants or the Chartered Institute of Management Accountants. Entry to their programmes is with three to five GCSEs/ S Grades (A-C/1-3), including English and maths, and two A-levels/ three H Grades. Degree entrants are exempt from certain of the examinations.

Management consultancy

Case Study

__Matthew__ did a Cambridge geography degree, after which he entered the surveying profession. After several years, he decided to broaden his employability by doing an MBA (Master in Business Administration) and now he works for a company of management consultants in the City. 'Besides a breadth of basic knowledge, geography has helped me to understand change in a number of contexts, to be able to disaggregate complex interactions into key elements and to be able to look at both the detail of something and its wider implications. I also think that the variety within geography prepared me very well for life as a consultant, for I work in a different industry on a different problem every few months.'

Management consultants give expert and sophisticated advice to all kinds and sizes of organization. The benefit of their perspective is an outside, and therefore objective, viewpoint. They are also capable of seeing a broader picture over a longer term. They advise particularly on how a company or organization manages change. The work of a management consultant is very varied, as every contract is different.

Getting started

It is rare to enter management consultancy directly from school, college or university. People who have had some extensive experience in the commercial or industrial world are required, for it is their expertise that counts. It may be seen as a second career rather than even a postgraduate one.

Information technology

Case Study

Danny did a geography degree in the north-west of England, after which he spent a year in each of accountancy and IT. He is now a telecommunications and IT consultant in the West End of London. 'Proposal writing every day and taking an analytical approach to problems come from geography, and it gave me the ability to understand people and "read between the lines" – important when you are managing a sales force.'

Case Study

Paul did geography on the south coast of England, before joining a major petroleum company as a graduate trainee. Seven years in the company brought him to the level of a product manager, and he has subsequently moved into IT. He is now a marketing systems manager. 'The material has not been particularly relevant but the skills and disciplines have: the ability to synthesize widely disparate data and generate meaningful conclusions, to solve problems creatively and to present ideas in a structured manner.'

The use of information technology in modern business is widespread. Larger organizations tend to manage their own IT systems, while smaller ones contract out to consultancies. Some specialist IT staff are also employed by companies manufacturing hardware, software and other IT equipment. Some IT specialists work for consultancies and some are self-employed.

Getting started

For the more technical and scientific aspects of IT, employers usually require graduates with degrees in more technical disciplines, like computing or maths. For programming they accept degrees or Higher National Awards related or unrelated to computing. Entrants to systems analysis will usually have a few years' experience of programming. Computing degrees require two or three A-levels/three to four H Grades, or BTec/SQA National Award or Advanced GNVQ/ GSVQ level 2.

Further reading

Taylor, Felicity (1996) *Careers in Accountancy*, Kogan Page, London

Longson, Sally (1999) *Careers in Banking and Finance*, Kogan Page, London

Yardley, David (1997) *Careers in Computing and Information Technology*, Kogan Page, London

Czerneda, Julie and Vincent, Victoria (1997) *Great Careers for People Interested in Communications Technology*, Kogan Page, London

14 Location, location and location

Some final thoughts

The preceding chapters have attempted to give a range of career ideas for students who have benefited from studying geography. Chapter 2 outlined what are seen as key skills that studying geography can develop. Chapters 12 and 13 took that theme a stage further by giving examples of taking those key skills into media and commercial jobs respectively.

The majority of chapters in the book, however, concentrated upon careers with a significant amount of the subject content in them (Chapter 1 generally and Chapters 3–11 specifically). Is there then a theme to unite these varied career areas? Well, geography is essentially about places, and the real focus of so much of the subject is location. When you think about it, geography teachers teach about locations (Chapter 3), planners plan different locations (Chapter 4), cartographers map them (Chapter 5), environment jobs attempt to protect them (Chapter 6), travel and tourism sell them for short stays (Chapter 7), development agencies aim to work in the poorest locations (Chapter 8), jobs in the physical environment focus on key processes in physical locations (Chapter 9), transport jobs essentially co-ordinate the transfer of people or goods from one location to another (Chapter 10), while an important component in analysing population is where people live (Chapter 11). Courses in geography then provide a locational springboard for a wide range of somewhat linked careers.

A particular focus for this locational emphasis in job advertisements is the region. Local government recruitment is at county level or below, and that involves a variety of departmental expertise. Marketing officers, media officers and inward investment officers essentially contribute to attracting investment to a region, thereby increasing jobs and prosperity in that place. The perspective of the job is geared to the region represented, whether it be broadening a rural economy, the regeneration of an urban-industrial one or marketing the heritage of a once-proud manufacturing region. All of these jobs reinforce the importance of the region to employment.

Are there new jobs with a geographical flavour emerging? Logistics is about controlling the supply chain of an industry or sector. In previous chapters, it emerged in two contexts: road haulage and the development sector. There are few logistics degrees available at present, but evidence in this book suggests that the overview required to appreciate each link in the supply chain is very much the synoptic view of the geographer. Indeed, the sort of diagram that geography teachers have traditionally applied to, for instance, the steel industry, with inputs, processes and outputs, seems to be just the methodology applied by logisticians. In addition, logisticians have to balance economic factors like pricing systems with social and political factors. More geographical methodology in the 'real world'!

The much maligned and often betrayed town centre has a new champion in the town centre officer or manager. Local authorities now realize that town centres have to be seen as entities, rather than as concentrations of shops, offices and services, if the problems they face are to be combated. Ironically, the major supermarkets, which are vital to a town centre's future, have had their own location-finding departments for almost 20 years. Indeed, in finding locations 'out of town' they have contributed to the decline of the traditional town centre, and to the need for a town centre manager! It would seem that this recently created job is ideal for a geographer. The context is familiar from the classroom and the lecture theatre, and the inter-personal skills and synoptic viewpoint of a good geographer can be put to good use.

As well as places as they are, there is also the fantasy of place. Some local authorities now lease locations for filming, and set up the arrangements for doing so, thereby complementing the people who act as location finders for film makers. Again, these are great examples of emerging geographical jobs.

The distinction between those jobs that are essentially 'geographical' and those that are not can, of course, mislead. You can work in a commercial career **and** within a geographical sector – a public relations officer in transport, for instance, or an accountant in the charity sector, or what about an environmental lawyer? Put your acquired commercial skills back into a geographical context.

Geography has been described as the perfect undergraduate degree because it takes a student from a broad base in the first year through increasing specialization thereafter. If you choose a geography degree, you can control its content once the foundation has been laid. That content can then determine what you choose to work in. It was interesting to hear the many people contacted as case studies noting differing advantages from their study of the subject. Whether they have remained in a traditionally geographical field or not, they could readily point to the transferable skills they have taken with them into many varied career fields. Geography is not just a great subject to study. Its sheer variety makes it a pivot, from which career paths can lead outwards in so many different directions.

Index

NB: page numbers in italics = tables, charts etc